my **revision** notes

WJEC GCSE

PHYSICS

Jeremy Pollard

HODDER
EDUCATION
AN HACHETTE UK COMPANY

Acknowledgements

Exam practice questions at the end of each chapter are reproduced by permission of WJEC.

Photo credits

p. 26 (top) © Washington Imaging/Alamy Stock Photo, (bottom) © ACORN 1/Alamy Stock Photo; p. 34 © sciencephotos/Alamy Stock Photo; p. 35 (left) © Berenice Abbott/Science Photo Library; (right) © Andrew Lambert Photography/Science Photo Library; p. 46 © Morrison1977 – iStock via Thinkstock/Getty Images

Although every effort has been made to ensure that website addresses are correct at time of going to press, Hodder Education cannot be held responsible for the content of any website mentioned. It is sometimes possible to find a relocated web page by typing in the address of the home page for a website in the URL window of your browser.

Orders: please contact Hachette UK Distribution, Hely Hutchinson Centre, Milton Road, Didcot, Oxfordshire, OX11 7HH. Telephone: +44 (0)1235 827827. Email education@hachette.co.uk. Lines are open from 9 a.m. to 5 p.m., Monday to Friday. You can also order through our website: www.hoddereducation.co.uk

Hachette UK's policy is to use papers that are natural, renewable and recyclable products and made from wood grown in well-managed forests and other controlled sources. The logging and manufacturing processes are expected to conform to the environmental regulations of the country of origin.

ISBN 978 1 4718 8356 9

First published in 2017 by
Hodder Education
an Hachette UK Company,
Carmelite House, 50 Victoria Embankment
London EC4Y 0DZ

Impression number   8

Year   2024

Cover photo © J.R. Bale/Alamy Stock photo

Typeset in Bembo Std Regular 11/13 by Integra Software Services Pvt. Ltd., Pondicherry, India

Printed by CPI Group (UK) Ltd, Croydon, CR0 4YY

A catalogue record for this title is available from the British Library

Get the most from this book

Everyone has to decide his or her own revision strategy, but it is essential to review your work, learn it and test your understanding. These Revision Notes will help you to do that in a planned way, topic by topic. Use this book as the cornerstone of your revision and don't hesitate to write in it — personalise your notes and check your progress by ticking off each section as you revise.

Tick to track your progress

Use the revision planner on pages iv–vi to plan your revision, topic by topic. Tick each box when you have:
- revised and understood a topic
- tested yourself
- practised the exam questions and gone online to check your answers and complete the quick quizzes

You can also keep track of your revision by ticking off each topic heading in the book. You may find it helpful to add your own notes as you work through each topic.

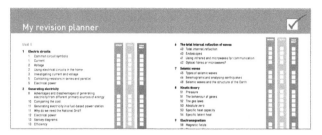

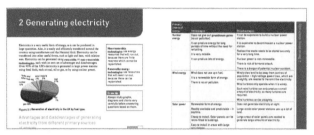

Features to help you succeed

Exam tips

Expert tips are given throughout the book to help you polish your exam technique in order to maximise your chances in the exam.

Now test yourself

These short, knowledge-based questions provide the first step in testing your learning. Answers are at the back of the book.

Definitions and key words

Clear, concise definitions of essential key terms are provided where they first appear.

Equations

The equations you should know how to use are given on page viii. A sheet of these equations is provided in the exam.

Exam practice

Practice exam questions are provided for each topic. Use them to consolidate your revision and practise your exam skills.

Summaries

The summaries provide a quick-check bullet list for each topic.

Online

Go online to check your answers to the exam questions and try out the extra quick quizzes at **www.hoddereducation.co.uk/myrevisionnotes downloads**

H ▶ Where this symbol appears, the text to the right of it relates to higher tier material.

My revision planner

REVISED TESTED EXAM READY

Exam practice answers and quick quizzes at
www.hoddereducation.co.uk/myrevisionnotesdownloads

Countdown to my exams

6–8 weeks to go

- Start by looking at the specification — make sure you know exactly what material you need to revise and the style of the examination. Use the revision planner on pages iv–vi to familiarise yourself with the topics.
- Organise your notes, making sure you have covered everything on the specification. The revision planner will help you to group your notes into topics.
- Work out a realistic revision plan that will allow you time for relaxation. Set aside days and times for all the subjects that you need to study, and stick to your timetable.
- Set yourself sensible targets. Break your revision down into focused sessions of around 40 minutes, divided by breaks. These Revision Notes organise the basic facts into short, memorable sections to make revising easier.

REVISED ☐

2–6 weeks to go

- Read through the relevant sections of this book and refer to the exam tips, summaries and key terms. Tick off the topics as you feel confident about them. Highlight those topics you find difficult and look at them again in detail.
- Test your understanding of each topic by working through the 'Now test yourself' questions in the book. Look up the answers at the back of the book.
- Make a note of any problem areas as you revise, and ask your teacher to go over these in class.
- Look at past papers. They are one of the best ways to revise and practise your exam skills. Write or prepare planned answers to the exam practice questions provided in this book. Check your answers online and try out the extra quick quizzes at **www.hoddereducation.co.uk/ myrevisionnotesdownloads**
- Try out different revision methods. For example, make notes using mind maps, spider diagrams or flash cards.
- Track your progress using the revision planner and give yourself a reward when you have achieved your target.

REVISED ☐

One week to go

- Try to fit in at least one more timed practice of an entire past paper and seek feedback from your teacher, comparing your work closely with the mark scheme.
- Check the revision planner to make sure you haven't missed out any topics. Brush up on any areas of difficulty by talking them over with a friend or getting help from your teacher.
- Attend any revision classes put on by your teacher. Remember, he or she is an expert at preparing people for examinations.

REVISED ☐

The day before the examination

- Flick through these Revision Notes for useful reminders, for example the exam tips, summaries and key terms.
- Check the time and place of your examination.
- Make sure you have everything you need — extra pens and pencils, tissues, a watch, bottled water, sweets.
- Allow some time to relax and have an early night to ensure you are fresh and alert for the examinations.

REVISED ☐

My exams

GCSE Physics Unit 1

Date:...

Time:...

Location:...

GCSE Physics Unit 2

Date:...

Time:...

Location:...

Equations

current = $\dfrac{\text{voltage}}{\text{resistance}}$	$I = \dfrac{V}{R}$
total resistance in a series circuit	$R = R_1 + R_2$
total resistance in a parallel circuit	$\dfrac{1}{R} = \dfrac{1}{R_1} = \dfrac{1}{R_2}$
energy transferred = power × time	$E = Pt$
power = voltage × current	$P = VI$
power = current² × resistance	$P = I^2 R$
% efficiency = $\dfrac{\text{energy [or power] usefully transferred}}{\text{total energy [or power] supplied}} \times 100$	
density = $\dfrac{\text{mass}}{\text{volume}}$	$\rho = \dfrac{m}{V}$
units used (kWh) = power(kW) × time (h) cost = units used × cost per unit	
wave speed = wavelength × frequency	$v = \lambda f$
speed = $\dfrac{\text{distance}}{\text{time}}$	
pressure = $\dfrac{\text{force}}{\text{area}}$	$p = \dfrac{F}{A}$
p = pressure, V = volume, T = kelvin temperature	$\dfrac{pV}{T}$ = constant
	$T/K = \theta/°C + 273$
change in thermal energy = mass × specific heat capacity × change in temperature	$\Delta Q = mc\Delta\theta$
thermal energy for a change of state = mass × specific latent heat	$Q = mL$
force on a conductor (at right angles to a magnetic field) carrying a current = magnetic field strength × current × length	$F = BIl$
V_1 = voltage across the primary coil, V_2 = voltage across the secondary coil, N_1 = number of turns on the primary coil, N_2 = number of turns on the secondary coil	$\dfrac{V_1}{V_2} = \dfrac{N_1}{N_2}$
acceleration [or deceleration] = $\dfrac{\text{change in velocity}}{\text{time}}$	$a = \dfrac{\Delta v}{t}$
acceleration = gradient of a velocity–time graph	
distance travelled = area under a velocity–time graph	
resultant force = mass × acceleration	$F = ma$
weight = mass × gravitational field strength	$W = mg$
work = force × distance	$W = Fd$
kinetic energy = $\dfrac{\text{mass} \times \text{velocity}^2}{2}$	$KE = \dfrac{1}{2}mv^2$
change in potential energy = mass × gravitational field strength × change in height	$PE = mgh$
force = spring constant × extension	$F = kx$
work done in stretching = area under a force–extension graph	$W = \dfrac{1}{2}Fx$
momentum = mass × velocity	$p = mv$
force = $\dfrac{\text{change in momentum}}{\text{time}}$	$F = \dfrac{\Delta p}{t}$
u = initial velocity v = final velocity t = time a = acceleration x = displacement	$v = u + at$ $x = \dfrac{u+v}{2}t$ $x = ut + \dfrac{1}{2}at^2$ $v^2 = u^2 + 2ax$
moment = force × distance	$M = Fd$

Exam practice answers and quick quizzes at **www.hoddereducation.co.uk/myrevisionnotes**

1 Electric circuits

This topic explores the relationship between current, potential difference (or voltage) and resistance. It shows how voltages and currents are related in series and parallel circuits, and how to calculate the total resistance a circuit. It looks at the concept of electrical **power** as the energy transferred per unit time and introduces the equations for the calculation of the power transferred by an appliance.

Power is the energy transferred per unit time.

Common circuit symbols

REVISED

Cell	—|⊢—	Ammeter	—(A)—
Battery	—|⊢···|⊢—	Voltmeter	—(V)—
Indicator lamp	—⊗—	Microphone	⊐◖
Filament lamp	—◯—	Bell	⌓
Switch	—o⁄o—	Buzzer	◁
Resistor	—▭—	Loudspeaker	◁◖
Variable resistor	—▱—	Motor	—(M)—
Diode	—▷|—	LED	▷|↗↗
Fuse	—▭—	LDR	↘↘▭
Thermistor	—◿—	Solar cell	↘↘⊕

Figure 1.1 Common circuit symbols.

Current

REVISED

The current flowing through electrical components in a circuit is measured in amperes (or amps), A, using an ammeter connected in series with the components.

Series circuits

For components connected in series, the current is the same at any point in the circuit. This means that all the components in a series circuit have the same current flowing through them. In Figure 1.2 the ammeter at A will read the same as an ammeter connected into the circuit at B or C.

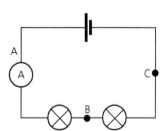

Figure 1.2 A series circuit.

Parallel circuits

When components are connected in parallel, the current splits when it gets to a junction in the circuit. No current is lost at a junction, so the total current into the junction equals the total current out of the junction. In Figure 1.3, the current at P is equal to the current at Q plus the current at R; the current at X plus the current at Y is equal to the current at Z.

> **Exam tip**
>
> Questions involving circuit diagrams require you to 'read' the diagram before attempting the question. Identify all the components and decide if the components are arranged in series or parallel. Pay particular attention to any labels next to components – they will have been put there for a reason!

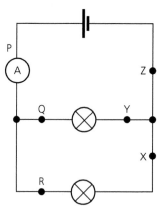

Figure 1.3 A parallel circuit.

Voltage

The voltage across components in a circuit is measured in volts using a voltmeter. Voltmeters are always connected in parallel across components, as in Figure 1.4. In series circuits, the voltages add up to the supply voltage. In parallel circuits, like Figure 1.3, the voltage is the same across each of the bulbs. A voltmeter connected between Q and Y will read the same voltage as one connected between R and X.

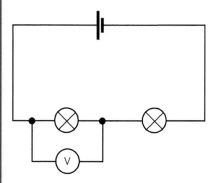

Figure 1.4 The voltmeter measures the voltage across one bulb.

Using electrical circuits in the home

Most of the mains electrical circuits in your house are connected in parallel. This has several advantages:

- If one component in the circuit stops working, all the others will continue to work properly.
- The voltage is the same for all the components.
- It is much easier to connect up all the circuits, and to add new circuits.
- It is easy to work out the total current being drawn by the different parts of the circuit (it all adds up).
- It is safer – each part of the circuit can be protected by its own fuse or circuit breaker and controlled by its own switch.

Now test yourself

1. How should an ammeter be connected into a circuit in order to measure a current?
2. A power supply provides 1.3A to a light bulb and 0.8A to a small electric motor connected in parallel with the light bulb. What is the total current drawn from the power supply?
3. What would be the advantage of connecting 12 Christmas tree lights in parallel, rather than in series?

Answers on page 119

Investigating current and voltage

Figure 1.5 shows a variable resistor connected in series with a fixed resistor. The resistance of the variable resistor can be changed, in order to vary the current through, and the voltage across, the fixed resistor. The fixed resistor could be replaced by any component, such as a filament lamp, to investigate how the current and voltage vary for the component.

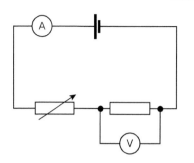

Figure 1.5 The variable resistor controls the current through, and voltage across, a fixed resistor.

Voltage–current relationships

The graphs in Figure 1.6 show the voltage–current relationships for a fixed resistor and a filament lamp.

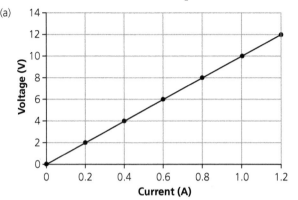

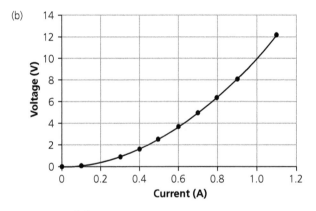

Figure 1.6 (a) Voltage against current for a 10 Ω fixed resistor and (b) a filament lamp.

- For fixed resistors (and wires at constant temperature), voltage and current are proportional to each other – doubling the current will double the voltage. The graph is linear (a straight line). A bigger resistance will give a bigger slope, on a *V–I* graph.
- For components such as filament lamps, the resistance changes with current. The resistance of a filament lamp increases with current, so the slope of the voltage–current graph increases.
- In the examination you could be shown *V–I* graphs (as in Figure 1.6) or *I–V* graphs, where current is plotted on the *y*-axis and *V* on the *x*-axis.

The current, voltage and resistance of electrical and electronic components are related to each other. The physicist Georg Ohm investigated this in 1827. We summarise his findings using the equation:

$$\text{current, } I \text{ (amps)} = \frac{\text{voltage, } V \text{ (volts)}}{\text{resistance, } R \text{ (ohms)}}$$

$$I = \frac{V}{R}$$

This equation can be used to calculate any one of the three variables, provided that we know the other two.

Examples

1 A $20\,\Omega$ (ohm) fixed resistor has a voltage of $12\,V$ across it. Calculate the current through it.
2 (Higher Tier) Calculate the resistance of a filament lamp operating at $6\,V$ with a current of $0.3\,A$ through it.

Answers

1 $I = \dfrac{V}{R} = \dfrac{12}{20} = 0.6\,A$

2 $I = \dfrac{V}{R}$ so $R = \dfrac{V}{I} = \dfrac{6}{0.3} = 20\,\Omega$

Thermistors, diodes and light-dependent resistors

Thermistors are components rather like resistors, but their resistance changes with temperature. Most thermistors decrease their resistance with temperature – these are called negative temperature coefficient (ntc) thermistors. Thermistors can be used in circuits as electrical temperature sensors.

Light-dependent resistors (LDRs) are components that change their resistance depending on the light intensity shining on them. They can be used as light sensors in electrical circuits. Most LDRs decrease their resistance with increasing light intensity.

Diodes are electrical components that control the direction of flow of the current in a circuit. They behave like one-way electrical gates, only allowing the current to flow in one direction through the diode. Figure 1.8 shows how the electrical characteristic graph for a diode could be determined, together with an example of an electrical characteristic.

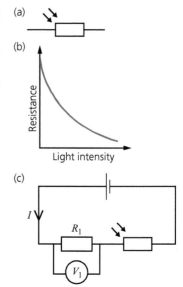

Figure 1.7 (a) The electrical symbol for an LDR. (b) The variation of the resistance of a typical LDR with light intensity. (c) An electrical circuit diagram showing how an LDR can be used in an electrical circuit.

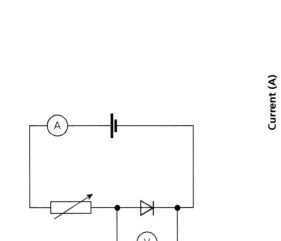

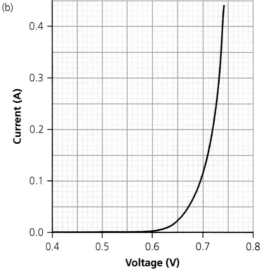

Figure 1.8 (a) A circuit for determining the I–V graph for a diode and (b) the electrical characteristic(I–V graph) for a diode.

Combining resistors in series and parallel

When two or more resistors (or components) are combined together in a series circuit (as shown Figure 1.9), the total resistance of the circuit increases and it is calculated from the sum of all the resistances using the equation:

$R = R_1 + R_2$

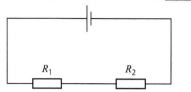

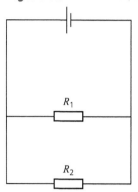

Figure 1.9 Resistors in series.

 Combining resistors in parallel reduces the overall resistance of the circuit. Figure 1.10 shows two resistors, R_1 and R_2, arranged in parallel with a battery.

The overall resistance, R, of a parallel circuit can be calculated using the equation:

$$\frac{1}{R} = \frac{1}{R_1} + \frac{1}{R_2} + \frac{1}{R_3} + \dots$$

Figure 1.10 Resistors in parallel.

Electrical power

The rate of transfer of electrical energy by a device is the electrical power, P, measured in watts, W. It can be calculated using the equation:

$P = VI$

or the equation:

$P = I^2R$

Examples

1   Calculate the power of a filament lamp operating at a voltage of 12V and a current of 0.5A.
2   A fixed resistor with a resistance of 25Ω has a current of 0.8A through it. Calculate the power of the fixed resistor.

Answers

1   $P = VI = 12 \times 0.5 = 6\,W$
2   $P = I^2R = 0.8^2 \times 25 = 16\,W$

Exam tip

Often an equation is given to you in an exam question, but sometimes you are asked to find a suitable equation from the formula sheet, which will be on the inside cover of the exam paper. On the Higher Tier paper, you may also need to rearrange an equation, so you need to practise this skill.

Now test yourself

4   The V–I graph for a filament bulb is shown in Figure 1.6 (b). Sketch the I–V graph for this bulb, with I on the y-axis and V on the x-axis.
5   Calculate the current flowing through a 15Ω resistor with a voltage of 3.0V across it.
6   Calculate the power of the resistor in Question 5.

Answers on page 119

Summary

- Common electrical circuit symbols are shown in Figure 1.1.
- In a series circuit, the current is the same throughout the circuit and the voltages add up to the supply voltage.
- In parallel circuits, the voltage is the same across each branch of the circuit and the sum of the currents in each branch is equal to the current in the supply.

- Voltmeters and ammeters are used to measure the voltage across, and current through, electrical components in electrical circuits.
- The voltage and current characteristics of a component can be shown on a V–I graph or an I–V graph.
- The equation for current I flowing through a component, where V is the voltage across a component of resistance, R, is

$$I = \frac{V}{R}$$

- Adding components in series increases total resistance in a circuit; adding components in parallel decreases total resistance in a circuit.

- The total resistance, R, of two resistors, R_1 and R_2, connected in series is given by:

$$R = R_1 + R_2$$

 ● The total resistance, R, of two resistors, R_1 and R_2, connected in parallel is given by:

$$\frac{1}{R} = \frac{1}{R_1} + \frac{1}{R_2}$$

- Electrical power is the electrical energy transferred per unit time:

$$E = Pt$$

- Electrical power can be calculated using these two equations:

$$P = VI$$

$$P = I^2R$$

Exam practice

1 Figure 1.11 shows part of a mains lighting circuit that is protected by a fuse in the household fuse box (consumer unit). A, B and C are lamps; S_1, S_2 and S_3 are switches.

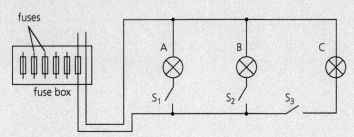

Figure 1.11

(a) Copy and complete the sentences below by selecting the correct words from the choice bracket. [2]
   If too much current is drawn by the lighting circuit, the fuse will melt. This makes the circuit [complete / incomplete] and the lamps will be [on / off].

(b) The fuse in this circuit is working properly. For a lamp to light there must be a complete circuit.
   (i) State which lamp(s) are lit when S_1 and S_2 are closed (on) and S_3 open (off). [1]
   (ii) State which lamp(s) are lit when S_3 is closed (on) and S_1 and S_2 open (off). [1]

WJEC GCSE Physics P2 Foundation Tier Summer 2010 Q9

2 Figure 1.12 shows part of a mains lighting circuit that is protected by a fuse in the mains fuse box (consumer unit). A, B, C and D are lamps in the circuit. The table gives information about each lamp.

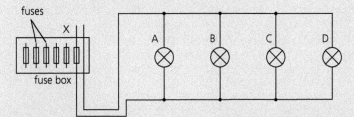

Lamp	Power (W)	Current (A)
A	40	0.17
B	60	0.26
C	40	0.17
D	60	0.26

Figure 1.12

(a) When working normally, calculate how much current is flowing through the fuse at X. [1]

(b) Add the following to the circuit diagram:
   (i) a switch labelled S_1 which controls lamp A only
   (ii) a switch labelled S_2 which controls lamps C and D only. [2]

WJEC GCSE Physics P2 Higher Tier Summer 2010 Q4

→

3 Figure 1.13 shows an ammeter A and a voltmeter V connected to a power supply and a resistance wire XY. A connector S allows the length of wire in the circuit to be changed.

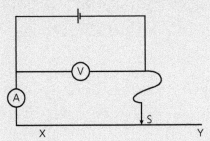

Figure 1.13

(a) With S in the position shown, the voltmeter reads 6 V and the ammeter 1.2 A. State a suitable equation that could be used to calculate the resistance of the wire between X and S, and then use the equation and the data to calculate this. [3]

(b) The connector S is moved towards Y. State the effect, if any, this would have on:

(i) the resistance in the circuit [1]

(ii) the ammeter reading. [1]

WJEC GCSE Physics P2 Higher Tier Summer 2010 Q1

4 The circuit shown in Figure 1.14 is used to investigate how the resistance of a lamp changes.

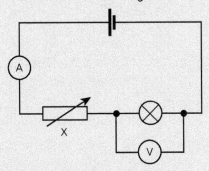

Figure 1.14

(a) Explain how component X allows a set of results to be obtained. [2]

(b) The results obtained are used to plot the graph shown in Figure 1.15.

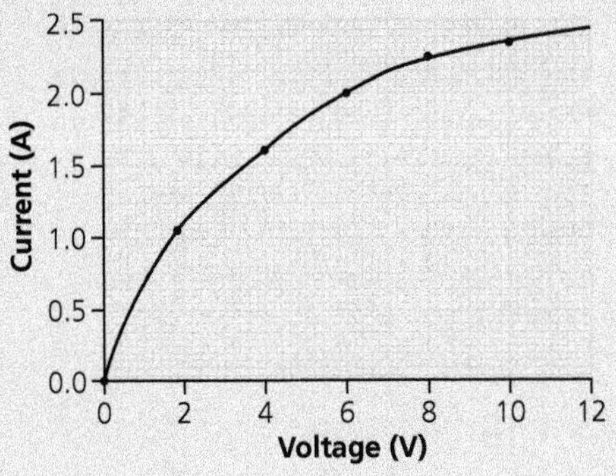

Figure 1.15

(i) Write down in words an equation from the equations list on page viii and use it to calculate the resistance of the lamp when the voltage across it is 4 V. [4]

(ii) Use the graph and a suitable equation from the equations list on page viii to calculate the power of the lamp when the voltage across it is 4 V. [3]

WJEC GCSE Physics P2 Higher Tier Summer 2008 Q6

Answers and quick quiz 1 online

ONLINE

2 Generating electricity

Electricity is a very useful form of energy, as it can be produced in large quantities. Also, it is easily and efficiently transferred around the country using transformers and the National Grid. Electricity can be transferred into other useful forms, such as light and heat, with relative ease. Electricity can be generated using **renewable** or **non-renewable technologies**, each with its own set of advantages and disadvantages. Over 90 per cent of the UK's electricity is generated in large power stations using fossil fuels, such as coal, oil or gas, or by using nuclear power.

Renewable energy technologies use resources that will never run out, because these can be replenished.

Non-renewable technologies use energy resources that will run out, because there are finite reserves which cannot be replenished.

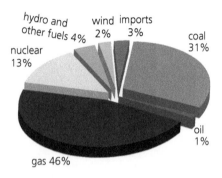

Figure 2.1 **Generation of electricity in the UK by fuel type.**

Exam tip

Always study graphs, diagrams and charts very carefully before answering questions based on them.

Advantages and disadvantages of generating electricity from different primary sources of energy

REVISED

Primary source of energy	Advantages	Disadvantages
Fossil fuels (such as coal, oil and gas)	Large amounts of electricity can be generated cheaply. Fossil-fuel power stations are very reliable. There is security of the supply of fossil fuels.	Fossil-fuel power stations can be dirty (especially coal). Burning fossil fuels produces carbon dioxide gas which contributes to the greenhouse effect and global warming. Burning fossil fuels produces sulfur dioxide gas, which contributes to acid rain. Large amounts of fuel need to be brought onto site and waste (in the case of coal) needs to be removed from the site. Fossil fuels are non-renewable forms of energy.

→

Primary source of energy	Advantages	Disadvantages
Nuclear energy	Does not give out **greenhouse gases** (no air pollution). It can produce energy for long periods of time without the need for refuelling. It is very reliable. It can produce lots of energy.	It can be expensive to build a nuclear power station. It is expensive to decommission a nuclear power station. Radioactive waste needs to be stored securely for a very long time. Nuclear power is non-renewable. There is risk of terrorist attack. There is a danger of potential nuclear accident.
Wind energy	Wind does not use up a fuel. It is a renewable form of energy. There is no air pollution.	Windy sites tend to be away from centres of population – high-voltage power lines, which are unsightly, are needed to transmit the electricity. Wind turbines only operate when it is windy. Each wind turbine can only produce a small amount of electricity, so many turbines are required. Wind turbines can be unsightly.
Solar power	Renewable form of energy. Readily available and predictable – in daytime. Cheap to install. Solar panels can be retro-fitted to buildings. Easy to install in areas with large populations.	Does not generate electricity at night. Large-scale solar power stations use up a lot of land. Large areas of solar panels are needed to generate large amounts of electricity.
Hydroelectric power (HEP)	HEP is renewable. There is no air pollution. Large HEP stations can generate enormous amounts of electricity reliably. HEP stations have almost instant start-up times, so can be switched on and off easily. There are no fuel costs. No fossil fuels are used.	Large dams need to be constructed, which can be expensive to build. Valleys are flooded when dams are constructed, destroying habitats. Suitable HEP sites tend to be away from centres of population – high-voltage power lines, which are unsightly, are needed to transmit the electricity. Drought can reduce the supply of water needed to produce HEP.
Wave and tidal energy	Both wave and tidal energy are renewable energies. Tidal energy is very predictable. Large-scale tidal power stations could generate huge amounts of electricity. No fossil fuels are used. Non-polluting. Both forms of power generation could be turned on and off very quickly.	Wave energy is unreliable and only works when there are suitable waves. Large numbers of wave generators would be needed to generate significant amounts of energy. Tidal energy barrages cause large-scale flooding of estuaries, destroying habitats.

→

Primary source of energy	Advantages	Disadvantages
Biofuels (such as animal waste, wood and fast-growing crops)	Biofuels are a renewable form of energy. Large-scale biofuel power stations could be built, generating large amounts of electricity.	Large areas of land are needed for fast-growing plants/trees, or a large amount of animal waste is needed, which would have to be transported cleanly. Although carbon neutral, carbon dioxide is still emitted into the atmosphere. Biofuel power stations can be unsightly.
Geothermal energy	Renewable form of energy. Does not produce any pollution. Reliable source of energy in places where there are hot springs or where hot rocks are close to the surface. Ground-source heat pumps can be installed into domestic houses. Cheap form of energy.	Hot springs and hot rocks are only available in certain areas, normally away from large populations, so unsightly pylons and cables are needed. Ground-source heat pumps need a lot of area to capture heat.

Exam tip

Extended writing questions require you to produce good quality written answers; you must be careful about the quality of the written communication that you use. Obvious things to check are: logical organisation of your thoughts and arguments; punctuation and grammar, such as correct use of capital letters and full stops; and correct use and spelling of key scientific terms.

A **greenhouse gas** is one that traps radiation in the Earth's atmosphere and, therefore, contributes to global warming. Carbon dioxide and methane are examples of greenhouse gases.

Comparing the cost

REVISED

Cost	Coal power station	Wind farm	Nuclear power station
Commissioning costs: Buying land Professional fees Building costs Labour costs	High	Low	Very high
Running costs: Labour costs Fuel costs	High	Very low	High
Decommissioning costs: Removal of fuel (nuclear) Demolition Clean-up	High	Low	Very high

Generating electricity in a fuel-based power station

REVISED

In a (fossil) fuel-based power station, the chemical energy stored within the coal, oil or gas is burnt in a furnace at several thousand degrees Celsius, together with air/oxygen, producing enough thermal heat energy to turn large amounts of water into steam every second in the boiler. The steam is then superheated and pressurised, causing it to have a huge amount of kinetic energy. This then turns turbines, spinning them at several thousand revolutions per minute (rpm). Each of the turbines is connected to an electric generator, producing large amounts of electrical energy, which is output to the National Grid. The superheated steam is then cooled and condensed back into water, by passing it through kilometres of pipes inside cooling towers.

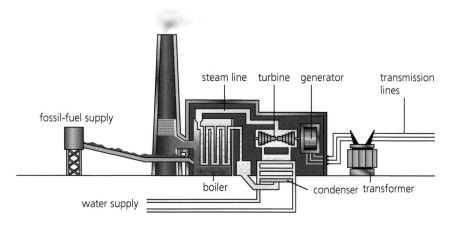

Figure 2.2 A schematic diagram of a typical fossil-fuel power station.

In a nuclear power station, a nuclear reactor is used to produce the heat needed to turn the water into superheated steam.

Why do we need the National Grid?

REVISED

Over 90 per cent of the UK's electricity is generated in large-scale power stations. The amount of electricity produced by these power stations is controlled by the National Grid, which provides:
● a reliable, secure energy supply
● an electricity supply that matches the changing demand during the day and over the course of the year
● high-voltage power lines that connect power stations to consumers
● electrical substations that control the voltage being supplied to consumers.

The amount of electricity consumed over the course of one day and over the course of the year varies in very predictable ways:

- Peak daily consumption is around 6 p.m., when people are cooking evening meals.
- Overall consumption is higher in the winter than in the summer, as people use more electricity for lighting and heating.

The National Grid

When electrical current passes down a wire, it causes the wire to heat up. The heat energy generated from the electricity then transfers into the surroundings, heating up the air. The larger the current, the greater the heat loss.

The National Grid is designed to minimise the amount of energy lost as heat when electricity passes down the power lines. The electricity generated at power stations is changed by step-up transformers to very high voltages (typically 400 000 V, 275 000 V or 132 000 V) but very low current – so that the energy lost as heat in the power lines is very small. (Only about 1 per cent of the total energy transmitted is lost in this way.)

High voltages would be very dangerous if used in homes and offices, so step-down transformers change the electricity to a lower voltage and higher current for use by consumers.

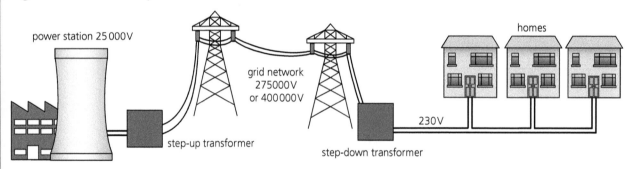

Figure 2.3 The National Grid transmission system.

Electrical power

Electrical power is a measure of the rate at which electrical energy can be transformed into other more useful forms of energy. Electrical power is calculated using the equation:

electrical power = voltage × current

$$P = VI$$

In most UK houses, mains voltage $V = 230\,\text{V}$.

Example

A mains hairdryer draws a current of 5.5 A. Calculate the power of the hairdryer.

Answer

Mains voltage = 230 V

Hairdryer current = 5.5 A

$$P = VI$$

power = 230 × 5.5 = 1265 W

Exam tip

When you are asked to do calculations involving units with prefixes (such as kV or MW) make sure that you convert the numbers carefully back into base numbers. For example, 400 kV = 400 000 V and 100 MW = 100 000 000 W.

Now test yourself

5 State the useful energy transfers within a fossil-fuelled power station.
6 Give two reasons why we need a National Grid.
7 Why is electricity transferred around the National Grid at high voltage and low current?
8 Calculate the power of a mains kettle operating at a voltage of 230V and a current of 13A.
9 Calculate the current flowing through a 2.5kW mains lawnmower operating at a voltage of 230V.

Answers on page 119

> **Exam tip**
>
> *Calculate* means that you must produce a numerical answer by doing a mathematical calculation.

Sankey diagrams

REVISED

The transfer of energy (or power) from one form to other forms can be shown using a Sankey diagram, which shows the types and amounts of energy as they transform into different forms. The Sankey diagram for an energy-efficient light bulb is shown in Figure 2.4.

Sankey diagrams are drawn to scale – the width of the arrow at any point shows the amount of energy being transformed. Conventionally, we write the type of energy and the amount of energy (or power) on the arrow, and the useful forms of energy usually go along the top of the diagram, with the wasted forms curving off downwards. Sankey diagrams give us not only a good way of showing energy (and power) transfers by a device or during a process, but also an indication of how efficient the process is – the bigger the useful energy arrow is compared to the input arrow, the higher the efficiency.

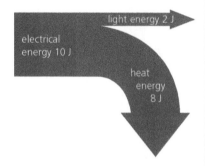

Figure 2.4 Sankey diagram for an energy-efficient light bulb.

Efficiency

REVISED

Efficiency is a measure of how much useful energy (or power) comes out of a device or process compared to the total amount of energy (or power) that goes into the device or process. Efficiency is usually expressed as a percentage using the equation:

$$\% \text{ efficiency} = \frac{\text{useful energy (or power) transfer}}{\text{total energy (or power) input}} \times 100$$

> **Example**
>
> The efficiency of an energy-efficient light bulb can be calculated using data from the Sankey diagram in Figure 2.4. The total energy input (as electricity) is 10J. The useful energy output (as light) is 2J.
>
> Answer
>
> $$\% \text{ efficiency} = \frac{\text{useful energy transfer}}{\text{total energy input}} \times 100$$
>
> $$= \frac{2}{10} \times 100 = 20\%$$

> **Exam tip**
>
> Foundation-tier students are not required to rearrange equations – but you might have to change numbers given with prefixed units, such as kW, back into base units, such as W.

Why is energy efficiency important?

Energy-efficient devices are very important for the future. The more efficient a device is, the more of the energy input is output as useful energy and less is wasted. Conventional fossil-fuel power stations are at best only 33 per cent efficient. This means that for every 100 tonnes of coal or oil used, only about 33 tonnes is converted directly into useful electricity. The rest of the coal or oil is effectively heating up the atmosphere and producing unnecessary carbon dioxide gas. Wind turbines are about 50 per cent efficient and solar panels about 30 per cent efficient. Standard tungsten filament light bulbs are typically only 2–3 per cent efficient; 'low-energy' bulbs are about 20 per cent efficient but LED light bulbs can be up to 90 per cent efficient. Imagine the effect on electricity consumption if every light bulb in the UK was replaced by an LED bulb.

Exam tip

You will come across two types of equation-based questions in the examination. You may be given an equation and asked to extract the correct data from the question to use in the equation; or you will be given the correct data and you will be asked to select the appropriate equation from the list at the front of the examination paper.

On the Higher Tier paper, you may also need to rearrange the equation as part of your answer.

Now test yourself

TESTED

10 Draw the Sankey diagram for 400 MW output gas-powered station, where 1000 MW of chemical energy is input from the gas.
11 Calculate the efficiency of the gas-powered power station in Question 10.
12 An 18 W LED light bulb is 90 per cent efficient. Calculate the power output as light.

Answers on page 119

Summary

- Electricity is a really useful form of energy because it is easy to produce and easy to transfer into other useful forms of energy.
- Power stations (renewable, non-renewable and nuclear) have significant but very different commissioning costs, running costs (including fuel) and decommissioning costs that need to be considered when planning a national energy strategy.
- Large-scale power generation by power stations and micro-generation using renewable technologies, e.g. using domestic wind turbines and roof-top photovoltaic cells, have different advantages and disadvantages. They all have very different environmental impacts.
- Data can be used to determine the efficiency and power output of power stations and renewable energies.
- In a fuel-based power station, electricity is generated by the burning fuel, producing thermal energy which boils water to produce steam. The moving steam turns a turbine, attached to a generator, which produces the electrical energy.

- Sankey diagrams can be used to show energy transfers.
- The efficiency of an energy transfer can be calculated from the equation:

$$\% \text{ efficiency} = \frac{\text{useful energy (or power) transfer}}{\text{total energy (or power) input}} \times 100$$

- There is a need for a national electricity distribution system (the National Grid), in order to maintain a reliable energy supply that is capable of responding to a fluctuating demand.
- The National Grid consists of power stations, substations and power lines.
- Electricity is transmitted across the country at high voltage because it is more efficient, but low voltage is used at home because it is safer.
- Transformers are needed to change the voltage and current within the National Grid.
- It is possible to experimentally investigate the operation of step-up and step-down transformers, in terms of the input and output voltage, current and power.
- power = voltage × current: $P = VI$

Exam practice

1 Study Figure 2.1 showing the proportions of electricity generated by different fuels. Calculate the percentage of UK electricity generated from fossil fuels and nuclear power combined. [1]

2 Discuss the factors that are involved in making decisions about the type of commercial power station that could be built in an area. [3]

WJEC GCSE Physics P1 Higher Tier Summer 2010 Q5(a)

3 Large coal-fired power stations are generally built close to lakes or rivers and near to both motorways and mainline railways. Suggest why coal power stations:
   (a) require good road and railway links [1]
   (b) are built near a source of water. [1]

WJEC GCSE Physics P1 Higher Tier January 2009 Q4(b)

4 If you live on the coast of Britain, the area may be ideal for building a power station nearby. The choice may be between building a nuclear or a coal-fired power station.
   (a) People often object to power stations because of their appearance. Write a paragraph describing three other objections you could raise to nuclear power stations. [3]
   (b) Write a paragraph describing three objections you could raise, apart from appearance, to coal-fired power stations. [3]

WJEC GCSE Physics P1 Higher Tier Summer 2008 Q6

5 The table shows some of the information that planners use to help them decide on the type of power station they will allow to be built.

	Wind	Nuclear
Overall cost of generating electricity (p/kWh)	5.4	2.8
Maximum power output (MW)	3.5	3600
Lifetime (years)	15	50
Waste produced	None	Radioactive substances, some of which remain dangerous for thousands of years
Lifetime carbon footprint (g of CO_2/kWh)	4.64 (onshore) 5.25 (offshore)	5

   (a) Give one reason why the information in the table does not support the idea that wind power will be a cheaper method of producing electricity. [1]
   (b) Supporters of wind power argue that it will reduce global warming more than nuclear power. Explain whether this is supported by information in the table. [2]
   (c) Supporters of nuclear power argue that it will meet a greater demand for electricity in the future than wind power. Give two ways in which this is supported by information in the table. [2]

WJEC GCSE Physics P1 Higher Tier January 2010 Q2

6 Figure 2.5 shows part of the National Grid. Electricity is generated at power station A.

Figure 2.5

   (a) Use the words below to copy and complete the sentences that follow. Each word may be used once, more than once or not at all.

   transformer   pylon   generator   power   current

   (i) At B, a _____ increases the voltage. [1]
   (ii) Electricity is sent at a high voltage along C, so the _____ is smaller. [1]
   (iii) At D, the voltage is decreased using a _____. [1]
   (b) Explain why the electricity is stepped up at B, but stepped down at D. [3]

(c) Assume that electricity is transmitted along the cables C at a power of 100 MW and a voltage of 400 kV. Use the equation: power = voltage × current to calculate the current in the cables. [3]

WJEC GCSE Physics P1 Foundation Tier January 2010 Q7

Exam tip

Read the instructions in questions carefully. Note that, in Question 6 (a) you can use any word from the list more than once and some words may not be used at all.

7 Figure 2.6 shows part of the National Grid.

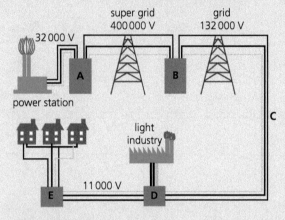

Figure 2.6

(a) At which point, A, B, C, D or E, would you find a step-up transformer? [1]
(b) What is the voltage at point C? [1]
(c) Where is the voltage stepped down to 230 V; give the letter A, B, C, D, or E? [1]
(d) Select the correct letter. A high voltage is used in the National Grid so that the electrical energy lost in the cables is:
  A zero
  B small
  C big. [1]

WJEC GCSE Physics P1 Foundation Tier January 2011 Q5

8 Water can be boiled using a saucepan on a gas-cooker ring. The energy transfers are shown in Figure 2.7.

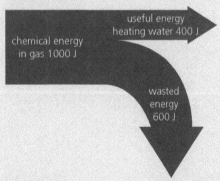

Figure 2.7

(a) Write down an equation and use it to find the efficiency of heating water in this way. [3]
(b) An electric kettle is 90 per cent efficient at boiling water. Copy and complete the energy transfer diagram in Figure 2.8. The diagram is not to scale. [2]

→

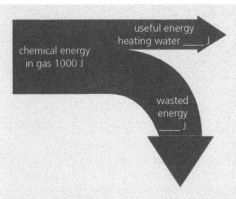

Figure 2.8

WJEC GCSE Physics P1 Higher Tier Summer 2010 Q3

9 The table shows how energy is used in a coal-burning power station. Write down in words a suitable equation and use it to calculate the efficiency of the power station. [3]

Energy input per second	Energy output per second
6000 MJ	3350 MJ of energy is taken away as heat in the water used for cooling
	2100 MJ of energy is fed into the National Grid
	550 MJ of energy is given out in the gases released during burning

WJEC GCSE Physics P1 Higher Tier January 2009 Q4(a)

Answers and quick quiz 2 online

ONLINE

3 Making use of energy

Conduction, convection and radiation

Homes are heated by transforming energy sources such as electricity or gas into heat using appliances such as electric fires or hot-water radiators. Thermal (heat) energy will move from somewhere hot (where the temperature is higher) to somewhere cold (where the temperature is lower). It does this by conduction, convection or radiation.

Conduction

Conduction is the transfer of energy from hot to cold by the successive vibration of particles within solids and liquids. Materials such as metals are very good thermal conductors because they have free mobile electrons within their structure. Materials that do not conduct thermal energy very well are called insulators – many non-metals are good insulators.

Convection

Convection occurs through liquids and gases. When a gas (or liquid) is heated, the particles move faster. As the particles speed up, they get further apart, increasing the volume of the gas. This causes the density of the gas to decrease. Less dense gas floats (or rises) above denser gas. As the gas rises it cools again, the particles slow down, get closer together and fall, increasing the density of the material. This creates a convection current which can heat a room. Temperature differences within the Earth's mantle and within the atmosphere cause natural convection currents.

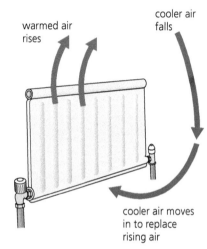

cooler air falls

warmed air rises

cooler air moves in to replace rising air

Figure 3.1 Convection currents transfer heat from the radiator to the room.

Radiation

Thermal radiation is emitted by hot objects. A hot-water radiator emits infrared electromagnetic radiation. Dull, black objects are good emitters and absorbers of thermal radiation. Shiny, light-coloured objects are good reflectors of thermal radiation. All objects emit thermal radiation, but the higher the temperature of the object the greater the amount of thermal radiation emitted.

Applying density

The density of a material can be calculated using the equation:

$$\text{density} = \frac{\text{mass}}{\text{volume}}$$

Solids have high densities because the particles inside them are generally closely packed together in a regular shape. Liquids have relatively high densities (but less than the corresponding solid) because the particles are still close packed (but free to move over each other). Gases have low densities as the particles are far apart.

Density is important when it comes to generating energy from several renewable sources such as wind energy.

Using wind energy

If:
- area of turbine blades is 25 m^2
- peak wind speed is 12 m/s
- density (of air) $= \dfrac{\text{mass (of air)}}{\text{volume (of air)}} = 1.2\,\text{kg/m}^3$

Then:

volume of air moving through turbine blades per second =

speed × area $= 12\,\text{m/s} \times 25\,\text{m}^2 = 300\,\text{m}^3$

mass of air moving through turbine per second =

density × volume $= 1.2\,\text{kg/m}^3 \times 300\,\text{m}^3 = 360\,\text{kg}$

kinetic energy of wind moving through turbine per second =

$\text{KE} = \dfrac{1}{2}mv^2 = 0.5 \times 360 \times 12^2 = 25\,920\,\text{J}$

And if the input power of wind is 25 920 W and the efficiency of the turbine is 25 per cent, then output power of turbine =

$25\,920\,\text{W} \times \dfrac{25}{100} = 6480\,\text{W} = 6.48\,\text{kW}$

Now test yourself

TESTED

1 Explain how thermal energy could be transferred away from a desk lamp.
2 Explain why metals are such good conductors of thermal energy.
3 Calculate the density of a 6 m^3 block of concrete that has a mass of 14 400 kg.
4 Calculate the mass of water per second that moves through a water turbine if 0.5 m^3 passes through per second and water has a density of 1000 kg/m^3.

Answers on page 119

Insulation

REVISED

The amount of thermal energy escaping from a house can be reduced by using domestic insulation systems that work by reducing the effects of **thermal conduction**, convection and radiation. The table summarises the main systems that can be installed.

Thermal conduction is the flow of heat energy through a material. Metals have very high conductivities. Brick and glass have lower conductivities, but energy still flows through them fast enough to cool a house down on a cold day.

The payback time for heating or insulation system is given by the equation:

$\text{payback time} = \dfrac{\text{installation cost}}{\text{annual savings}}$

Insulation system	How it works	Typical installation costs	Typical annual savings	Payback time (years)
Draught proofing	Draught excluders and draught-proofing strips are fitted, reducing the convection of hot air through gaps under doors and in window frames.	£50	£50	1
Cavity-wall insulation	Fills the space between the double walls of bricks with foam. The foam traps air, which is a poor conductor and prevents air from circulating within the cavity, reducing thermal loss by convection.	£250	£110	2.3
Floor insulation	Mineral wool is laid between the joists under the floorboards and silicone sealant is used to seal gaps between skirting boards and floorboards. This reduces thermal loss via conduction and convection.	£140	£70	2
Loft insulation	Mineral wool insulation is laid between the timber joists in the loft. This reduces thermal loss via conduction and convection.	£250	£150	1.7
Double glazing	Two sheets of glass with a gap between them. Reduces thermal loss via conduction and convection.	£2000	£130	15.4

Units of energy – the kilowatt-hour

REVISED

Home energy values are usually compared in units equivalent to the **kilowatt-hour**, kWh.

1 kWh is equal to the amount of heat energy produced by a 1 kilowatt (1000 W) electric fire in one hour (3600 s).

$1\,kWh = 1000\,W \times 3600\,s = 3\,600\,000\,J$

> The **kilowatt-hour** is a unit of energy used in the domestic context.

Heating and transport costs

REVISED

Homes can be heated using a variety of different fuels. The table summarises some of the costs involved.

Fuel	Fuel price (p per unit)	Unit	Cost per kWh of fuel (p)	Energy content (kWh per unit)	CO_2 emissions per kWh
Electricity	16.8	kWh	16.8	1.0	0.5
Gas	4.7	kWh	5.2	1.0	0.2
Oil	54.1	litre	5.8	10.4	0.3
LPG (liquid propane gas)	36.7	litre	6.1	6.7	0.2
Butane	137.0	litre	19.1	8.0	0.2
Propane	74.2	litre	11.7	7.1	0.2
Wood	20.8	kg	5.8	4.2	0.03
Coal	30.0	kg	5.8	6.9	0.4

The costs associated with transport are quite variable, as they are very dependent on the world wholesale energy prices. A simple comparison table compares three variations of the Renault Clio/Zoe car; one diesel, one petrol and the electric Zoe.

Car (fuel)	Cost to buy	Road tax	CO_2 emissions (g/km)	Average fuel cost per year
Renault Clio Expression+ TCe ECO (petrol)	£13 245	£120	99	£1028
Renault Clio Expression+ dCi ECO (diesel)	£14 345	£100	83	£740
Renault Zoe Expression	£13 650	Zero	Zero	£1006 (including battery rental)

Now test yourself

TESTED

5 Which thermal energy transfer process is reduced by fitting draught excluders?
6 A house is heated with 54 MJ of thermal energy. Calculate the number of kWh of thermal energy used.
7 A builder is researching which fuel would be best for a new-build house. Use the table of data on page 20 to decide which fuel would give the cheapest running costs.
8 The builder in Question 7 wants to install cavity-wall insulation into the new-build house. The insulation will cost £350 to install and the insulation manufacturer has quoted a payback time of 2.5 years. Calculate the typical annual savings of this system.

Answers on page 119

Summary

- Temperature differences lead to the transfer of energy thermally by conduction, convection and radiation.
- The density of an object is given by the equation:

$$density = \frac{mass}{volume}$$

- Differences in density occur between the three states of matter, due to the different arrangements of the atoms or molecules.
- Thermal conduction happens from hot to cold areas in solids and liquids due to vibrations being passed from particle to particle. Metals are particularly good conductors of thermal energy due to the presence of mobile electrons within their structure.
- Thermal convection happens in liquids and gases. Hot particles move faster, get further

apart, increase the volume of the material and decrease its density. This causes the hotter particles to float above the colder denser material, forming a convection current.
- Energy loss from houses can be restricted by systems such as: loft insulation; double glazing; cavity-wall insulation and draught excluders.
- The cost effectiveness and efficiency of different methods of reducing energy loss from the home can be compared in order to test their effectiveness. This can include calculating the payback time and the economic and environmental issues surrounding controlling energy loss.
- Data can be used to investigate the cost of using a variety of energy sources for heating and transport.

Exam practice

1 A homeowner decided to reduce their heating bill by improving their house insulation. The table below shows the cost of the improvements made and the yearly savings.

Insulation method	Cost	Yearly saving
Draught-proofing doors and windows	£80	£30
Fitting a jacket to the hot-water tank	£20	£20
Cavity-wall insulation	£1100	£50
Loft insulation	£400	(i)
Total	(ii)	£200

(a) Copy and complete the table by giving values for (i) and (ii). [2]

(b) The homeowner spent £1200 per year heating his house before insulating it. How much would he expect to spend each year after the improvements? [1]

(c) Give a reason why heat loss by convection is reduced by cavity-wall insulation. [1]

WJEC GCSE Physics P1 Foundation Tier Summer 2010 Q4

2 (a) State how double glazing reduces the amount of heat lost through the windows of a house. [2]

(b) The graph in Figure 3.2 shows the results of an investigation to see how the rate of loss of energy through a double-glazed window was affected by the width of the air gap between the two panes of glass. The investigation used a window of area 1 m² and kept a temperature difference of 20 °C between the inside and the outside.

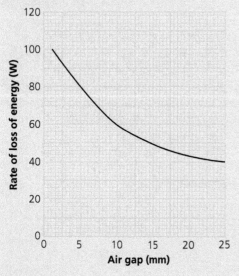

Figure 3.2

(i) Use the graph to estimate the rate of loss of energy for an air gap of 0 mm, and explain how you obtained your answer. [2]

→

(ii) Give two reasons why most manufacturers of double-glazed windows are unlikely to use an air gap any larger than 20 mm. [2]

WJEC GCSE Physics P1 Higher Tier Summer 2009 Q4

3 A water turbine is sited in a river flowing at 2 m/s. The density of water is 1000 kg/m³ and 0.15 m³ of water passes through the turbine per second.

(a) Calculate the mass of water flowing through the turbine per second. [2]

(b) The water turbine produces an electrical output of 48 W. The water inputs 120 W of kinetic energy. Calculate the efficiency of the water turbine. [1]

4 A householder buys gas for heating and cooking, and electricity for lighting and operating electrical appliances. The table shows information about the householder's energy consumption and the total yearly cost.

Year	Units of electricity (kWh)	Units of gas (kWh)	Total units of energy (kWh)	Total cost (£)
1st Jan–31st Dec 2015	4309	36958	41267	866.62
1st Jan–31st Dec 2016	4540	33446	37986	949.65

(a) Use the data from the table to find the overall cost of 1 unit (kWh) of energy in 2016. [3]

(b) On 1st January 2016 the householder fitted a solar panel, at a cost of £2000, to provide hot water for heating.

(i) Use data from the table to estimate the number of units produced by the solar panel in 2016. [1]

(ii) Use the answer from part (a) to calculate the amount of money he saved on his 2016 gas bill. [1]

(iii) Calculate the time it would take for his annual savings to pay back the cost of the solar panel. [2]

(iv) Give a reason why the payback time calculated in (iii) could be much smaller. [1]

WJEC GCSE Physics P1 Higher Tier January 2009 Q7

Answers and quick quiz 3 online

ONLINE

4 Domestic electricity

How much does it cost to run?

REVISED

The cost of running an electrical device depends on the electrical power of the device (given in kW, where 1 kW = 1000 W) and the electricity tariff (the cost per unit of electrical energy), and is calculated as follows:

energy transfer = power × time

or

$E = P \times t$

The cost of domestic electricity energy consumed is calculated using kilowatt-hours (kWh) or units. 1 kWh is equivalent to the amount of electrical energy used by a standard 1 kW electric fire in 1 hour. Units of electrical energy are calculated using the equation:

units used (kWh) = power (kW) × time (h)

The cost of the electrical energy is then given by:

cost = units used × cost per unit

The power rating of an electrical appliance is written on a plate attached to the appliance, and most domestic appliances in the UK are sold with an energy banding value (A–G), which tells you how efficient the appliance is. This is either written on the box that the appliance was sold in, or is attached as a sticker or a tag.

> **Exam tip**
>
> Be careful when calculating electrical energy consumption in kilowatt-hours. The time must be in hours. If you are given a time in minutes, you must convert it to hours first.

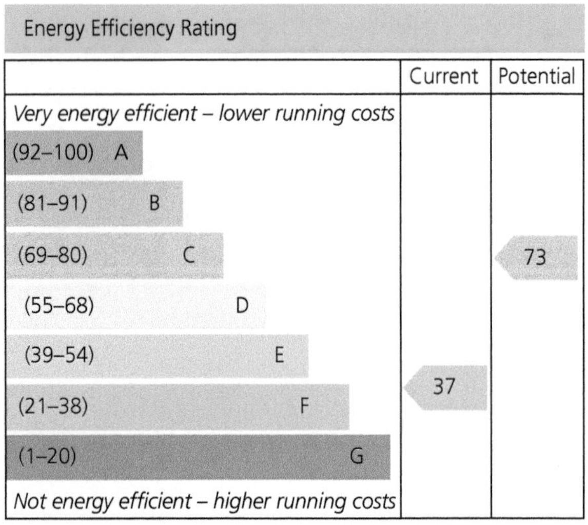

Energy Efficiency Rating		
	Current	Potential
Very energy efficient – lower running costs		
(92–100) A		
(81–91) B		
(69–80) C		73
(55–68) D		
(39–54) E		
(21–38) F	37	
(1–20) G		
Not energy efficient – higher running costs		

Figure 4.1 Energy efficiency rating label for an electrical appliance.

Now test yourself

1 A 200 W lamp is left on for 15 minutes. How much electrical energy is transferred?
2 A 2 kW heater is left on in your room. You put it on at 6 a.m. and forget about it until 5 p.m. If a unit (1 kWh) costs 15p, how much will it have cost to leave the heater on?

Answers on page 119

a.c. or d.c.?

Electricity can either be direct current, d.c., whereby the electric current flows only in one direction, or it can be a.c., alternating current, whereby the electric current flows in one direction for half a cycle, then in the other direction for the rest of its cycle. An oscilloscope can be used to show the difference between the two types of current. Example traces are shown in Figure 4.2.

Power supplies such as batteries and solar cells produce d.c. and generators produce a.c. In the UK, a.c. current has a frequency of 50 Hz.

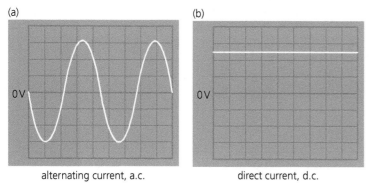

(a) alternating current, a.c.

(b) direct current, d.c.

Figure 4.2 Oscilloscope traces for (a) alternating current and (b) direct current.

Using mains electricity safely

Domestic electricity is supplied to your house at 230 V with a maximum current of about 65 A. The distribution of the electric current around your home is controlled by a consumer unit, consisting of the supply connection and a series of circuit breakers that control the amount of electric current being supplied to each individual circuit around the house.

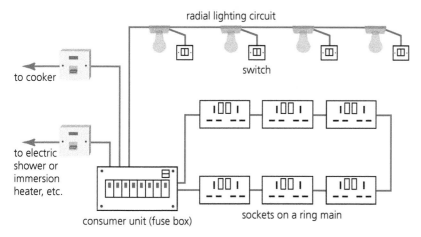

Figure 4.3 A typical domestic electricity connection system.

The consumer unit not only distributes the electric current around the house, it is also a safety system, preventing too much electric current from being drawn from the supply and cutting off any circuits that are faulty (short circuit). Consumer units are a series of circuit breakers, consisting of miniature circuit breakers (mcb) and residual current circuit breakers (rccb), as shown in Figure 4.4.

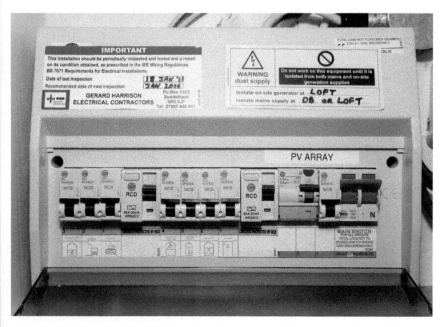

Figure 4.4 An electricity consumer unit.

Mcb switches control individual circuits within the house, such as the lighting circuits. They allow circuits to be turned on or off, and also limit the amount of current being drawn. If a short circuit occurs, the current being drawn from the consumer unit rises very quickly, exceeding the value of the mcb current rating. This turns off the circuit, isolating it from the supply and making the circuit safe. Rccb switches monitor the difference between the current drawn from the consumer unit and the current returning to it. If the difference between the two currents exceeds the rating on the rccb, it switches off, isolating the circuit from the supply. Rccb switches reduce the risk of electrocution to the people in the house because, if someone accidently touches the live wire of any part of the circuit, a larger current will be drawn from the consumer unit than returns to it, triggering the rccb and breaking the circuit. Mcb and rccb switches can be reset.

> **Exam tip**
>
> Mcb and rccb switches both isolate the electricity supply. Rccb switches work very quickly so they reduce the risk of electrocution.

Fuses

Fuses are like circuit breakers in appliances in that, if the current exceeds the rating (or maximum current) of the fuse, the wire within the fuse quickly heats up and melts, disconnecting the appliance from the supply. Fuses are fitted to the live wire of the plug and they prevent too much current from flowing through an appliance – which could cause a fire. Unlike mcb switches, standard cartridge fuses cannot be reset; they have to be replaced.

Figure 4.5 A standard 13A fuse and its circuit symbol.

The ring main

The main socket circuit within a house is called the ring main. Only one cable is needed to connect all the sockets in the house.

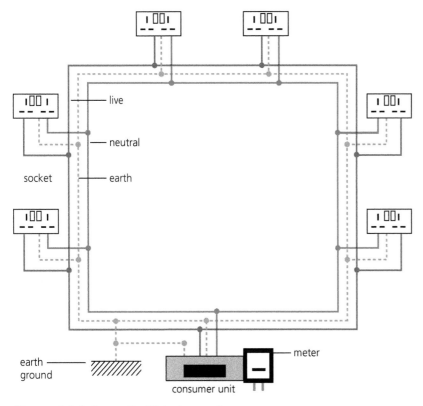

Figure 4.6 A domestic 32 A ring main.

The ring main cable consists of three wires: the live wire (always coloured brown), which carries the current from the consumer unit to the sockets; the neutral wire (always coloured blue), which returns the current to the consumer unit; and the earth wire (always coloured green/yellow), which acts as a safety system for the circuit. The earth wire connects the material of the whole installation to an earth connection at the local electricity sub-station. If the live wire becomes loose and creates a short circuit, the electric current flows down the earth wire safely into the ground, reducing the risk of electrocution.

Now test yourself

TESTED

3 Calculate the current normally drawn from a 2.2 kW mains kettle, and use this value to decide on the rating of the fuse that should be fitted to its plug.
4 What is the difference between a.c. and d.c.?
5 Explain how an electricity consumer unit is used as a safety system for a house.
6 What is the difference between an mcb and a cartridge fuse?

Answers on page 120

Micro-generation of domestic electricity

REVISED

Micro-generation of electricity means generating electricity locally on a small scale and close to where it is needed. Examples of micro-generation are roof-top photovoltaic cells and domestic wind turbines. Micro-generation has many advantages and some disadvantages over the large-scale generation of electricity from power stations.

Advantages of micro-generation

- Does not produce carbon dioxide and so does not contribute towards the greenhouse effect and global warming.
- Does not produce sulfur dioxide or oxides of nitrogen and so does not contribute towards acid rain.
- Zero fuel costs.
- Higher efficiency of generation.
- Can sell some electricity back to National Grid (feed-in).
- Roof-top photovoltaics:
  ○ provide 'free' electricity during daylight hours
  ○ on average, can generate 3 kW of electricity (peak).
- Domestic wind turbines:
  ○ provide 'free' electricity when the wind is blowing
  ○ on average, can generate 6 kW of electricity (peak).

Disadvantages of micro-generation

- Erratic energy supply.
- Can have long payback times.
- Cannot generate large quantities of electricity in one place.
- Many locations are very limited in which types of micro-generation can be used.
- Some people object to the visual impact of wind turbines and solar panels.
- Roof-top photovoltaics:
  ○ have a visual impact on roof-tops
  ○ need to cover a large area to generate large amounts of electricity.
- Domestic wind turbines:
  ○ cause a visual impact from the turbine
  ○ cause an impact from the noise of the turbine
  ○ are unsuitable for most locations as they need an exposed windy site.

> **Exam tip**
>
> A common extended writing question involves comparing micro-generation systems, such as wind turbines and roof-top photovoltaics, with more conventional methods of generating electricity, such as gas-fired power stations. Remember to compare the advantages and disadvantages of both methods.
>
> Remember to use capital letters and full stops in your answer, and spell the keywords correctly.

Now test yourself

TESTED

7 A roof-top wind turbine costs £3500 to install and will save £700 of electricity costs per year. Calculate the payback time of the turbine.

8 Wind turbines and roof-top photovoltaics both produce 'free' energy. Why are photovoltaics less erratic than wind turbines?

Answers on page 120

Summary

- The kilowatt (kW) is a convenient unit of power used by domestic appliances. The kilowatt-hour (kWh) is the unit of energy used by electricity companies when charging customers.
- The cost of electricity can be calculated using the equations:

  units used (kWh) = power (kW) × time (h)

  cost = units used × cost per unit

- The power of a domestic appliance can be obtained either directly from its power rating plate or through the energy banding (A–G) sticker.

- Electric currents can be either alternating current (a.c.) or direct current (d.c.). Direct currents only flow in one direction, whereas alternating current flows in one direction for half a cycle, then in the opposite direction for the rest of the cycle.
- Fuses, miniature circuit breakers (mcb) and residual current circuit breakers (rccb) are devices in mains electrical circuits (and appliances) to limit the current flowing through the circuit, making the circuits safer. The rating of a fuse is the maximum current that can flow through it before the special wire within the fuse

→

melts, disconnecting the circuit. The rating of the fuse fitted in a plug is always a little higher than the standard operating current of the appliance.
● A domestic ring main is a way of connecting up the sockets within a house. The live wire carries the electricity from the consumer unit, the neutral wire returns it to the consumer unit and the earth wire acts as a safety system in case there is a short circuit.

● The cost effectiveness of introducing domestic solar and wind energy equipment into a house is determined by the installation cost of the equipment and the fuel cost savings. The payback time is the amount of time (in years) that the equipment needs to be fitted for before the savings start to outweigh the installation costs.

Exam practice

1 Fuses and circuit breakers are electrical safety devices used to protect household electrical circuits.
   (a) Explain how fuses and miniature circuit breakers protect household electrical circuits. [2]
   (b) State one way in which miniature circuit breakers are more effective than fuses. [1]
   (c) Explain how the action of a residual current device is different from that of a miniature circuit breaker. [2]

   WJEC GCSE Physics P2 Higher Tier Summer 2007 Q6

2 Circuits and users are protected by the following safety features:
   fuse   miniature circuit breaker (mcb)   residual current circuit breaker (rccb)   earth wire
   (a) Name a safety feature that prevents cables becoming too hot. [1]
   (b) Name a safety feature that detects a difference in current between the live and neutral wires. [1]
   (c) Name a safety feature that will cause a fuse to blow if current flows through it. [1]
   (d) Name a safety feature that needs replacing once it acts. [1]

   WJEC GCSE Physics P1 Foundation Tier Summer 2008 Q1

3 The lighting circuit in a house is protected by a 5A fuse and connected to 230V. The table shows the current taken by different lamps.

Power of lamp (W)	Current (A)
40	0.17
60	
100	0.43

   (a) Use the equation below to find the current through a 60W lamp. [1]

   $$current = \frac{power}{voltage}$$

   (b) The circuit shows three lamps in a household lighting circuit connected to a 5A fuse. We can calculate the current through the fuse by adding up the currents through each of the lamps. Use the information in the table to find the current flowing through the fuse when all these lamps are switched on. [2]
   (c) Find the maximum number of 100W lamps that could be connected in a 5A household lighting circuit. [2]

   WJEC GCSE Physics P2 Higher Tier January 2008 Q1

4 The table shows information about three electrical appliances.

Appliance	Power (W)	Power (kW)	Units (kWh) used in 1 week
Kettle	2100	2.1	5
Electric oven		4.0	12
Microwave oven	900	0.9	1

   (a) (i) What does 'kW' stand for? [1]
       (ii) Complete the table. [1]
       (iii) State which appliance uses the most energy every second. [1]

(b) Calculate the number of hours that the electric oven is used in 1 week. [1]

(c) All three appliances are used for 1 week.

   (i) Calculate the total number of units used. [1]

   (ii) If 1 unit of electricity costs 12p. Calculate the cost of using all three appliances for 1 week. [2]

WJEC GCSE Physics P1 Foundation Tier January 2011 Q7

5 40% of all the wind energy in Europe blows over the UK, making it an ideal country for small home wind turbines. Roof-mounted turbines produce around 1 kW to 2 kW depending on wind speed. To be effective you need an average wind speed bigger than 5 m/s. Small domestic wind systems are particularly suitable for use in remote locations where homes are not connected to the National Grid. Costs for a roof-mounted wind system are £1500. Recent monitoring of a range of small domestic wind systems has shown that a well-sited 2 kW turbine could save around £300 a year off electricity bills.

(a) Why is the UK ideal for small home wind turbines? [1]

(b) The average wind speed in one town is 3.5 m/s. Give a reason why homeowners here would not be advised to install wind turbines. [1]

(c) Why are wind turbines useful for supplying electricity to farms on hilltops well away from towns? [2]

(d) Calculate the payback time for the roof-mounted wind turbine mentioned in the passage. [1]

WJEC GCSE Physics P1 Foundation Tier Summer 2010 Q5

Answers and quick quiz 4 online

ONLINE

5 Features of waves

Transverse and longitudinal waves

There are two types of waves: **transverse waves** (like water waves), in which the direction of motion of the wave is at right angles to the direction of vibration of the wave, and **longitudinal waves** (like sound waves), in which the direction of motion is in the same direction as the direction of vibration of the wave. Transverse waves travel as a series of peaks and troughs; longitudinal waves travel as compressions and rarefactions. Both of these sorts of waves can be demonstrated with a slinky spring as shown in Figure 5.1.

> A **transverse wave** is one in which the vibrations causing a wave are at right angles to the direction of energy transfer.
>
> A **longitudinal wave** is one in which the vibration causing the wave is parallel to the direction of energy transfer.

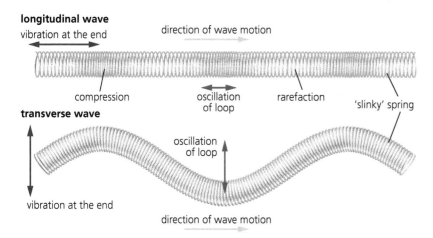

Figure 5.1 Longitudinal and transverse waves on a slinky.

How do we describe waves?

Waves are described in terms of their **wavelength**, **frequency**, **speed** and **amplitude**. Figure 5.2 shows these quantities on a transverse wave, like a water wave or light wave, in which the direction of vibration of the wave is at right angles to the direction of travel of the wave.

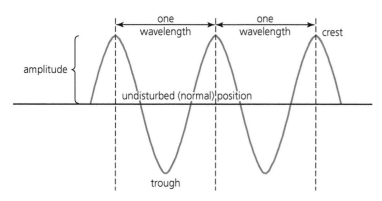

Figure 5.2 Wave measurements.

> The **wavelength**, λ, of a wave is the distance that the wave takes to repeat itself – this is normally measured from one crest to the next crest. Wavelength is measured in metres, m.

> The **frequency**, f, of a wave is the number of waves that pass a point in 1 second. Frequency is measured in hertz, Hz, where 1 Hz = 1 wave per second.
>
> The **speed** of a wave, v, is the distance that a wave travels in 1 second. Wave speed is measured in metres per second, m/s.
>
> The **amplitude** of a wave is a measure of the energy carried by the wave. Amplitude is measured from the undisturbed (normal) position to the top of a crest or the bottom of a trough. Loud sounds have larger amplitudes than quiet sounds.

Calculating wave speed, frequency and wavelength

REVISED

- Wave speed can be calculated using the equation:

$$\text{wave speed} = \frac{\text{distance}}{\text{time}}$$

- Wave speed, frequency and wavelength are all related by the basic wave equation:

$$\text{wave speed} = \text{frequency} \times \text{wavelength}$$

$$v = f\lambda$$

- Waves travel at a range of different speeds.
- All electromagnetic waves travel at the speed of light, $c = 300\,000\,000\,\text{m/s}$ or $3 \times 10^8\,\text{m/s}$.
- Water waves, like surf, travel at about $4\,\text{m/s}$.

Now test yourself

TESTED

1 A surfer takes 10 s to travel 50 m on the crest of a wave onto a beach. What is her speed?
2 The wavelength of the waves in Question 1 is 40 m. What is the frequency of the waves?
3 Calculate the speed of sound waves travelling through wood with a frequency of 5 kHz and a wavelength of 79.2 cm.
4 Explain the difference between a transverse and a longitudinal wave.

Answers on page 120

What is the electromagnetic spectrum?

REVISED

The electromagnetic spectrum is a family of (transverse) waves that all travel at the same speed in a vacuum, $300\,000\,000\,\text{m/s}$ or $3 \times 10^8\,\text{m/s}$. Like the energy given out by radioactive materials, electromagnetic waves are also called 'radiation'.

> **Radiation** refers to electromagnetic waves or the energy given out by radioactive materials.

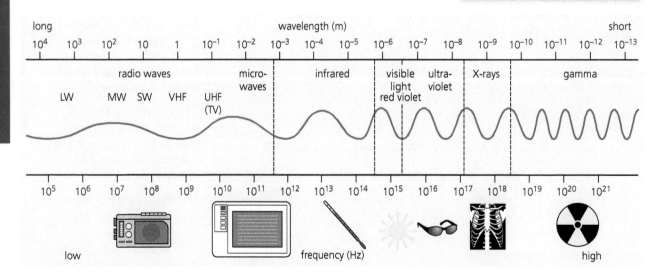

Figure 5.3 **The electromagnetic spectrum.**

The different parts of the electromagnetic spectrum have different wavelengths and frequencies, and energies. The higher the frequency of a wave, the higher its energy. The frequency, wavelength and energy of an electromagnetic wave completely determine its properties and how it will behave.

- Gamma rays have very high energy and can ionise (kill or damage) cancer cells, but they are also used to make images of the body.
- X-rays are also ionising, and are also used in medical imaging.
- Ultraviolet light can ionise skin cells causing sunburn.

The short wavelength parts of the electromagnetic spectrum (ultraviolet, X-rays and gamma rays) are all called ionising radiations, because they are able to interact with atoms and damage cells due to their large energies.

- Infrared radiation is used for heating and communications, such as in TV remote controls and optic fibres.
- Microwaves are also used for heating and communicating, particularly as mobile phone signals.
- Radio waves are used for communications over much longer distances, transmitting TV and radio programmes.

All parts of the electromagnetic spectrum can carry information and energy. Stars also emit all parts of the spectrum, giving us information about their composition and behaviour. Visible light, infrared, microwaves and radio waves are commonly used by human beings to transmit information.

Now test yourself

TESTED

5 What is meant by ionising radiation?
6 Which part(s) of the electromagnetic spectrum:
  (a) has the lowest energy
  (b) has a frequency range between visible light and X-rays
  (c) can be used for cooking
  (d) are emitted by stars
  (e) are used for human communications?

Answers on page 120

Reflection of waves

REVISED

Reflection is a fundamental property of all waves. When straight (plane) wave-fronts hit a flat barrier, they rebound off, obeying the law of reflection. Figure 5.4 shows this happening with water waves in a ripple tank.

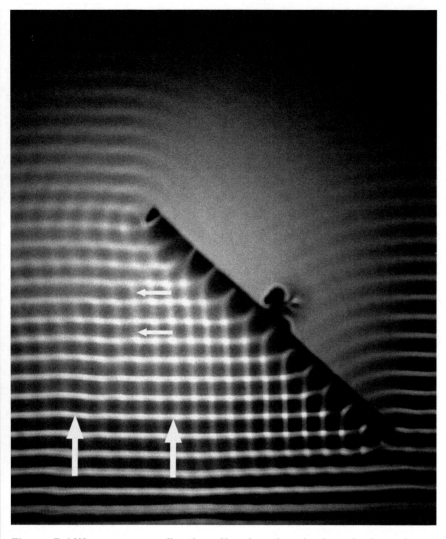

Figure 5.4 Water waves reflecting off a plane barrier in a ripple tank.

Figure 5.5 explains the reflection of waves.

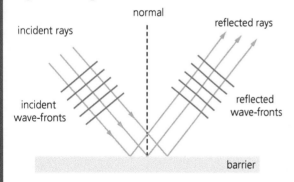

Figure 5.5

> The **normal** is a line drawn at 90° to a surface where waves are incident.
>
> The **angle of incidence** is the angle between the incident ray and the normal.
>
> The **angle of reflection** is the angle between the reflected ray and the normal.

The imaginary rays, drawn at right angles to the wave-fronts, show the direction of travel of the wave-fronts. The angles between the incident and reflected rays and the **normal** line (an imaginary line at right angles to the barrier/mirror) are equal, obeying the law of reflection, where:

angle of incidence = angle of reflection

Refraction of waves

REVISED

When water waves travel from deep water into shallow water, they slow down and the wave-fronts get closer together, decreasing their wavelength. This effect is called **refraction**. When the wave-fronts hit the boundary between the deeper water and the shallow water at an angle, they appear to change direction as shown in Figure 5.6.

Figure 5.6 The refraction of water waves in a ripple tank.

The refraction of light through a glass block in Figure 5.7 shows the rays of light changing direction as they go from the air into the glass, and then back again.

Figure 5.7 The refraction of light through a glass block.

Figure 5.8 is a diagrammatic version of Figure 5.7, with the angles labelled.

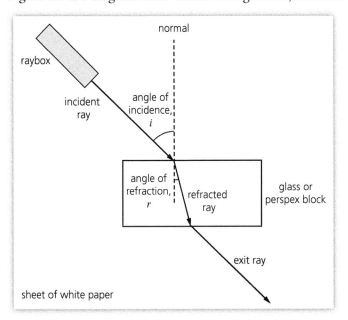

Figure 5.8 The angle of incidence and the angle of refraction.

Communicating using satellites

Mobile phones use microwaves and, as they are wireless signals, you don't need a copper cable or an optical fibre to transmit them. However, one disadvantage of using microwaves is that there must be a clear path between the transmitter and the receiver, which might be your television aerial or mobile phone. To cover the largest area, TV and mobile phone transmitters are tall and sited on hills. The curvature of the Earth means that repeater stations have to relay the microwave signal to distant transmitters. Satellites must be used for long-distance communications around the world. Theoretically, only three satellites are needed to transmit signals around the world. In practice, more are used. The satellites are placed in orbit at a height of 36 000 km. They orbit the Earth, above the equator, exactly in time with the Earth's rotation. This is called a geosynchronous (geostationary) orbit.

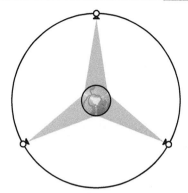

Figure 5.9 The Earth from above the North Pole; three geosynchronous satellites could send signals to the Earth.

Now test yourself

TESTED

7 What is the law of reflection?
8 What is the 'normal' line?
9 Explain why the wavelength of water waves decrease as they travel from deep water to shallow water.
10 Why are three satellites needed for long-distance microwave communications around the world?

Answers on page 120

Summary

- Transverse waves vibrate at right angles to their direction of motion of the wave. Longitudinal waves vibrate in the same direction as the direction of motion.
- Waves can be distinguished in terms of their wavelength, frequency, speed, amplitude (and energy).
- The equations associated with waves are:

$$\text{wave speed} = \frac{\text{distance}}{\text{time}}$$

wave speed (m/s) = frequency (Hz) × wavelength (m)

- Waves will reflect when they hit a barrier, obeying the law of reflection.
- Waves will refract when they go across a boundary between one medium where they travel at one speed into another medium where they travel at a different speed. Refraction causes the wavelength of the wave to change.

- All regions of the electromagnetic spectrum transmit information and energy.
- The electromagnetic spectrum is a continuous spectrum of waves of different wavelengths and frequencies consisting of radio waves, microwaves, infrared, visible light, ultraviolet radiation, X-rays and gamma rays, but all the waves travel at the same speed in a vacuum – the speed of light.
- The term 'radiation' can be used to describe both electromagnetic waves and the energy given out by radioactive materials.
- Radioactive emissions and the short wavelength parts of the electromagnetic spectrum (ultraviolet, X-ray and gamma ray) are ionising radiations, and they can interact with atoms, damaging cells by the energy that they carry.
- Microwaves and infrared radiation are used for mobile phones, intercontinental optical-fibre links, and for long-distance communication, via geosynchronous satellites.

Exam practice

1 The graph in Figure 5.10 shows how the frequency of deep ocean waves depends on the wavelength of the waves.

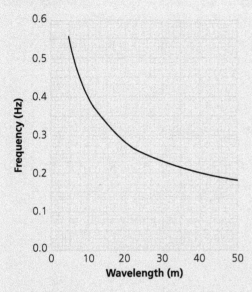

Figure 5.10

(a) Use information from the graph in Figure 5.10 and the equation below to calculate the speed of waves with a wavelength of 40 m.   [2]

wave speed = wavelength × frequency

(b) A large meteorite falls into the ocean and produces waves with a range of wavelengths.
    (i) Use the equation below to calculate how long it would take 40 m-wavelength waves to arrive at an island 5600 m away.   [1]

$$speed = \frac{distance}{time}$$

    (ii) Would 10 m waves arrive before or after the 40 m waves? Use information from the graph to explain your answer.   [2]

WJEC GCSE Physics P1 Foundation Tier Summer 2009 Q7

2 Figure 5.11 shows a train of waves.

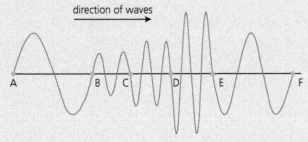

Figure 5.11

(a) How many waves are shown between A and C? [1]

(b) Between which two of the points, A–F, is:
   (i) the wavelength biggest [1]
   (ii) the amplitude smallest? [1]

(c) The eight waves between A and F cover a distance of 240 cm. Calculate the average wavelength of the waves. [1]

WJEC GCSE Physics P1 Higher Tier January 2011 Q1

3 Yellow light travels to us from the Sun at a speed of 3×10^8 m/s. It has a frequency of 5×10^{14} Hz. Write down in words a suitable equation and use it to calculate the wavelength of this yellow light. [3]

WJEC GCSE Physics P1 Higher Tier Summer 2008 Q5(a)

4 (a) Using the words below fill in the missing parts of the electromagnetic spectrum, (i) and (ii). [2]

   ultraviolet   radio waves   sound waves   water waves

(i)	Microwaves	Infrared	Visible light	(ii)	X-rays	Gamma rays

(b) Some electromagnetic waves can be used for communications.
   (i) Name the wave that is used by remote controls. [1]
   (ii) Name the wave that is used to communicate with satellites in space. [1]

(c) Some of these waves can be harmful.
   (i) Name one wave from the list that can ionise cells in the body. [1]
   (ii) What is the danger from a large dose of infrared rays? [1]

WJEC GCSE Physics P1 Foundation Tier January 2011 Q6

5 Figure 5.12 shows a communications satellite A in geosynchronous (geostationary) orbit around the Earth. The diagram is not to scale.

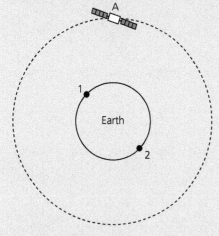

Figure 5.12

(a) (i) Explain the advantages of placing communications satellites in geosynchronous orbit. [2]
   (ii) Copy the diagram and add another satellite B and repeater station 3 that will enable radio station 1 to communicate with radio station 2. [2]
   (iii) Show on the diagram the path taken by the signal, via the satellites A and B, when radio station 1 communicates with radio station 2. [1]

(b) (i) Communications between geosynchronous satellites and Earth are made using microwaves of wavelength 20 cm that travel at 3×10^8 m/s. Use a suitable equation to calculate the frequency of the microwaves. [3]
   (ii) The time delay between sending a signal from 1 and its reception at 2 is 0.48 s. Use a suitable equation to find the approximate height of geostationary satellites above the Earth. [3]

WJEC GCSE Physics P1 Higher Tier January 2010 Q6

→

6 (a) Electromagnetic waves are used in communications to send television signals.
  (i)   Name the part of the spectrum that carries TV signals via satellites.   [1]
  (ii) Name the part of the spectrum that carries TV signals from transmitters to a home TV aerial.   [1]
  (iii) Name a part of the spectrum that carries TV signals through optical-fibre cables.   [1]
 (b) A householder installs a dish to receive TV signals from a communication satellite. Explain why
     the householder will not need to move the dish once it is set up.   [2]

WJEC GCSE Physics P1 Foundation Tier Summer 2010 Q8

7 Figure 5.13 shows light travelling from air into a glass block.

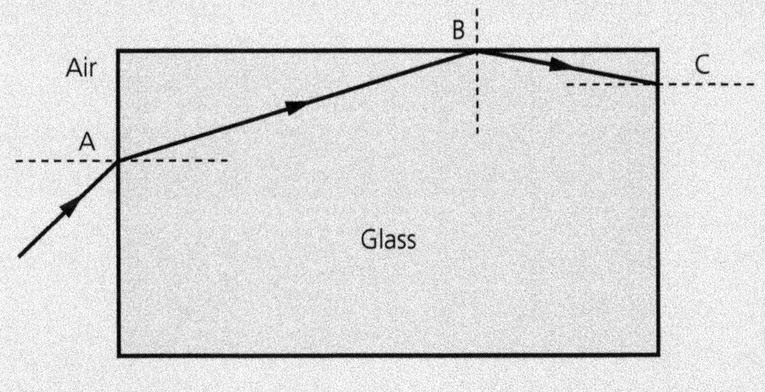

Figure 5.13

(a) (i) What name is given to the bending of light at point A?   [1]
    (ii) Give a reason why the light changes direction at A.   [1]
(b) State what law is being obeyed at point B.   [1]
(c) At point C, the light passes out into the air.
    (i) Give one reason why it does not go back into the block as it does at point B.   [1]
    (ii) Draw the ray direction into the air at point C.   [1]
(d) A long and very thin glass block becomes an optical fibre. Name a type of the electromagnetic
    radiation (other than visible light) that can be used to send messages along an optical fibre.   [1]
(e) The speed of signals along optical fibres is 2.0×10^8 m/s. Select an equation and use it to find
    the time that a signal would take to travel from London to New York along an optical fibre if the
    distance is 4.8×10^7 m. Give the correct unit for your answer.   [3]

WJEC GCSE Physics P1 Higher Tier January 2011 Q3

Answers and quick quiz 5 online

ONLINE

6 The total internal reflection of waves

Optical fibres make superfast internet connections and keyhole surgery possible. They consist of long, thin flexible tubes of glass, surrounded by a coating that allows a beam of light or infrared ray to continually reflect down the fibre. This is called total internal reflection.

Total internal reflection

Figure 6.1 shows what happens when a beam of light is passed through a glass block. The angle i is the angle of incidence, measured from the normal. For angles of i with low value, the beam refracts through the back face of the block, as shown by the red line. The angle of refraction increases until an angle of incidence called the critical angle is reached (about 42° for glass), and the angle of refraction is 90°. This is shown by the blue line in the diagram. For angles greater than the critical angle, the beam is reflected back into the block, obeying the law of reflection.

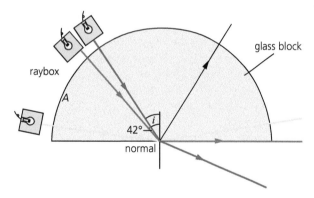

Figure 6.1 Refraction and reflection in a glass block.

> **Exam tip**
>
> The critical angle of total internal reflection depends on the materials involved. Water to air has a critical angle of about 50°, glass to air is about 42° and polycarbonate plastic to air is about 39°.

Endoscopes

Optical fibres are very thin and flexible, making them ideal for use in medical endoscopes. A medical endoscope generally has two sets of optical fibres. One set **transmits** light from a source down through the endoscope and another set picks up the light that is reflected from the inside of the body, transmitting it back up the scope, so an image can be displayed on a screen.

> A medium that **transmits** light is one that allows light to pass through it.

Endoscopes make 'keyhole' surgery possible. They can be inserted into the body through the mouth or the anus, to access the digestive system; other parts of the body can be accessed by making a small keyhole-sized incision in the skin, through which the endoscope tube is passed into the body cavity or the blood stream. The advantages of using endoscopy are:
- Recovery times for patients are very quick, and infections are kept to a minimum.
- No ionising radiation is used, reducing the chances of damage to non-affected cells.
- Biopsies (small tissue samples) can be taken by a probe at the end of the endoscope, allowing cell and tissue samples to be analysed.

● Real-time, close-up, colour images or video of internal body features can be viewed.

Now test yourself

TESTED ☐

1 Explain what happens when a beam of light hits a glass–air boundary with an angle of incidence greater than the critical angle.
2 Why does an endoscope need two sets of optical fibres?
3 State two advantages of using an endoscope, rather than an X-ray, to determine if a patient has a cancerous growth near their liver.

Answers on page 120

Using infrared and microwaves for communication

REVISED ☐

Real-time, long-distance communications are possible because of satellite and optical-fibre links, as shown in Figure 6.2.

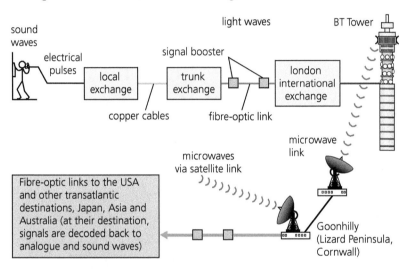

Figure 6.2 The path of an international phone call as it leaves the country.

Mobile phones connect to a world-wide network of microwave and optical-fibre links, transmitting phone signals across the globe and back at the speed of light. Land–line calls travel down optical fibres, using infrared radiation. Optical fibres are better at transmitting information than the copper wires that used to carry long-distance phone calls; a single optical fibre can carry over 1.5 million telephone conversations (compared with 1000 conversations for copper wires) or ten television channels. An intercontinental optical cable carries many fibres, so huge amounts of information can be transferred cost-effectively.

For long-distance telephone calls, electrical signals are converted to digital pulses. The digital signal is then converted to light pulses by an infrared laser that flashes at high speed. Repeaters boost the signal at 30 km intervals along the fibre. At the far end, a decoder converts the digital signal from the laser into a changing voltage, which is then converted into sound at the telephone earpiece.

Other advantages of optical fibres over copper wire are:
● Fibre optic lines use less energy.
● They need fewer boosters.
● There is no cross-talk (interference) with adjoining cables.
● They are difficult to bug.
● Their weight is lower, so they are easier to install.

Optical fibres or microwaves?

Both optical fibres and satellite communications are used for international phone calls and TV broadcasts. It takes time for the signals to travel from an Earth station up to one of the satellites and back again, as shown in Figure 6.3. This generates a time delay in the signal.

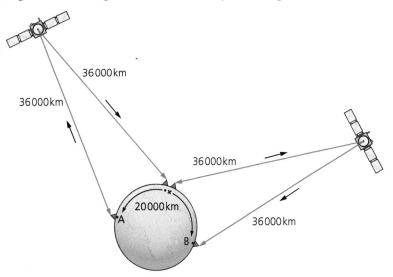

Figure 6.3 The satellite signal has much further to travel.

The satellites orbit at a height of 36 000 km so the path length is 4 × 36 000 km, or 144 000 km. Use the formula below to calculate the time delay for studio-to-studio transmission via satellite:

$$\text{time taken (s)} = \frac{\text{distance travelled (km)}}{\text{speed (km/s)}}$$

$$\text{time taken (s)} = \frac{144\,000\ \text{km}}{300\,000\ \text{km/s}} = 0.5\ \text{s (approx)}$$

An outside broadcast could increase the signal's journey to 200 000 km, making the time delay about 0.7 s. This time delay is quite noticeable on news broadcasts and telephone conversations.

With optical fibres connecting two studios, the distance travelled may be only 20 000 km, and infrared waves travel at 200 000 km/s in optical fibres:

$$\text{time delay (s)} = \frac{20\,000\ \text{km}}{200\,000\ \text{km/s}} = 0.1\ \text{s (approx)}$$

So, the time delay with optical fibres is only 0.1 s, which is much less noticeable.

> **Exam tip**
>
> Comparing the time delays between signals sent as microwaves via satellite and signals sent as infrared via optical fibres is a very common question in examination papers. Remember that microwaves travel there and back via satellites, so the distance travelled is generally *twice* the distance to the satellite. The speed of infrared inside an optic fibre is 200 000 000 m/s, *not* the speed of light in air.

Now test yourself

4 Which type of electromagnetic radiation is used to:
  (a) connect mobile phones to their base aerials
  (b) pass signals down optical-fibre communication links
  (c) connect ground stations to satellites?
5 Explain why signals transmitted via optical fibres travel faster than those sent via copper cables.
6 Why do signals travelling down optical fibres need to be repeated every 30 km?
7 The surface distance from Cardiff to Auckland, New Zealand, is about 18 400 km. Calculate the time delay in a video link between these two cities:
  (a) via an optical-fibre link, where the infrared signals travel at 200 000 000 m/s
  (b) via a geostationary satellite in orbit above the equator, 38 000 000 m from Cardiff and 38 000 000 m from Auckland. Microwaves travel at 300 000 000 m/s through the atmosphere.

Answers on page 120

Summary

- Light (and other forms of waves) will undergo total internal reflection if the light crosses a boundary from a medium where it is travelling slowly into a medium where it is travelling faster, at an angle of incidence greater than the critical angle for the boundary.
- Endoscopes and optical fibres rely on total internal reflection for their operation.
- Optical fibres using infrared waves and geosynchronous satellites using radio waves or microwaves can be used for long-distance communication. Optical fibres can carry a large number of signals and have shorter time delays than satellite communications, but require a fixed connection unlike satellite communications.
- Optical fibres can be used for endoscopic medical examinations. Endoscopes produce high-quality, close-up, real-time images and biopsies can be taken. Endoscopy is non-ionising and does not damage non-affected cells, unlike CT scans.

Exam practice

Exam tip

In many calculation questions, you get a mark just for identifying and using a suitable equation. A list of equations will be given to you on a datasheet in the examination paper.

1 A geosynchronous (geostationary) satellite is used to send signals from A to B.

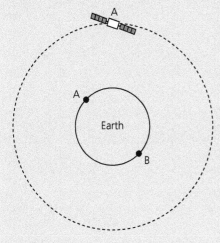

Figure 6.4

(a) (i) How long does it take this satellite to orbit the Earth once? [1]
   (ii) Copy Figure 6.4 and add to it to show how A sends signals to B via the satellite. [1]
(b) Explain why communications satellites are put into geostationary orbits. [1]
(c) A microwave signal, of frequency 5×10^9 Hz and speed 3×10^8 m/s, carries TV pictures from a studio to a geostationary satellite 3.6×10^7 m above the equator. The satellite receives the signal and then transmits it back to Earth, where it is received by homes with satellite dishes.
   (i) Use a suitable equation to calculate the wavelength of the microwave signal. [3]
   (ii) Use a suitable equation to calculate the time for the TV pictures to travel from the studio to the homes of viewers. [3]

WJEC GCSE Physics P1 Foundation Tier January 2010 Q5

2 This question concerns long-distance communications between two distant points on the Earth's surface, A and B, using satellite and optical-fibre links.
(a) Information is passed from A to B using a satellite in geosynchronous orbit 3.6×10^4 km above the Earth's surface. Microwaves carry the information at a speed of 3×10^8 m/s from A to B via the satellite. Use a suitable equation to calculate the time delay between sending and receiving the information. [4]

→

(b) (i) The information could also be sent from A to B via a transcontinental optical fibre linking A and B. An infrared signal carries the information at a speed of 2×10^8 m/s. Give a reason why the time delay between sending and receiving this signal is much shorter than that calculated in part (a). [1]

(ii) State two other advantages of using optical fibres to send information over long distances. [2]

WJEC GCSE Physics P1 Higher Tier Summer 2009 Q5

3 Figure 6.5 shows what happens to a beam of light as it emerges from a glass block.

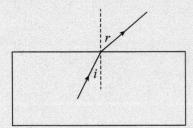

Figure 6.5

(a) (i) Name this effect. [1]
(ii) Give a reason why the beam changes direction. [1]
(b) When angle i is 42°, angle r is 90°. Show this on a copy of Figure 6.6. [1]

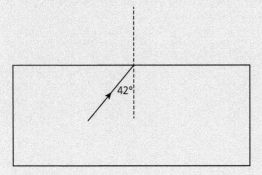

Figure 6.6

(c) (i) Copy and complete the diagram to show what happens to the beam of light if angle i is 50°. [1]

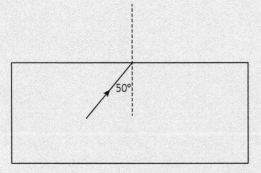

Figure 6.7

(ii) What is the name given to this effect? [1]
(iii) Hence, state two conditions needed for this effect to occur in optical fibres. [2]

(d) Infrared radiation of frequency 4×10^{13} Hz has a wavelength of 5×10^{-6} m in a glass fibre.
(i) Use a suitable equation to calculate the speed of the infrared radiation in the glass fibre. [3]
(ii) Use a suitable equation to calculate the time taken by an infrared signal to travel along a glass fibre 10 km long. [4]

WJEC GCSE Physics P1 Higher Tier Summer 2007 Q3

Answers and quick quiz 6 online

ONLINE

7 Seismic waves

Earthquakes happen when tectonic plates move relative to each other, releasing huge stresses that generate seismic waves. The effects of the seismic waves at the Earth's surface cause the earthquake, producing damage or generating powerful tsunamis.

Types of seismic waves

Earthquakes generate three types of seismic wave and the two main types are shown in Figure 7.1.

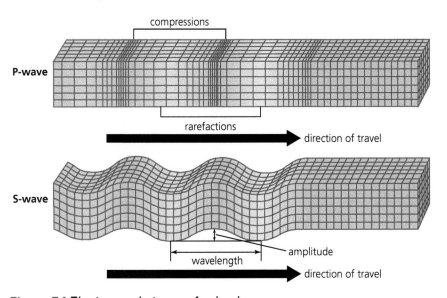

Figure 7.1 The two main types of seismic wave.

Primary (**P-waves**) are longitudinal waves where the direction of vibration of the rock is in the same direction as the propagation of the wave. P-waves always arrive first after an earthquake as they travel fastest (approximately 5–8 km/s). P-waves are produced by a push–pull motion of the rock, so they can travel through both solid and liquid rock and can be detected all over the Earth's surface following an earthquake.

> **P-waves** are longitudinal seismic waves.
>
> **S-waves** are transverse seismic waves.

Secondary (**S-waves**) are transverse waves where the direction of vibration of the rock is at right angles to the direction of propagation of the wave. S-waves are slower (approximately 2–8 km/s) and are produced by a shear motion of the rock; as such, they cannot travel through liquids. The liquid outer core of the Earth, therefore, forms a shadow region on the opposite surface of the Earth from an earthquake where only P-waves are detected.

The third type of seismic wave is surface waves. These propagate more slowly across the surface of tectonic plates (typically with speeds of between 1 and 6 km/s).

> **Exam tip**
>
> P-waves are primary waves, meaning first, so they travel fastest and are detected first. S-waves, being transverse waves, have a shape of a letter S on its side.

Now test yourself

1 Which type of seismic wave travels as transverse waves?
2 Why do S-waves always arrive second after an earthquake?
3 Which type of seismic wave is generated by a push–pull motion of the rock particles?
4 Why do S-waves form a shadow region on the opposite side of the Earth from an earthquake?
5 Which type of seismic wave can move through the liquid outer core of the Earth?
6 What are surface waves?

Answers on page 120

Seismograms and analysing earthquakes

Earthquakes and seismic waves can be detected by a seismometer and recorded on a seismogram, which is a visual record of the vibrations of the Earth caused by an earthquake. Seismometers work because they have a large mass attached to sensors that produce a small current as the mass and the sensors move relative to each other when seismic waves pass. The small current is recorded by a computer and displayed as a seismogram trace on a screen.

Figure 7.2 A seismogram following an earthquake.

On a seismogram, the passing of time is always represented on the trace from left to right and the amplitude of the vibrations is shown vertically on the trace. The faster P-waves will always be shown first on the trace, followed by the S-waves and then the surface waves. An important measurement on a seismogram is the time lag between the arrival of the P-waves and the arrival of the S-waves. This value can be used to determine the distance from the seismometer to the epicentre of the earthquake. Three of these measurements from different seismic stations can be used to triangulate and determine the position of the epicentre.

Analysing an example seismic trace

Following an earthquake, seismograms from two different monitoring stations, A and B, are shown in Figure 7.3.

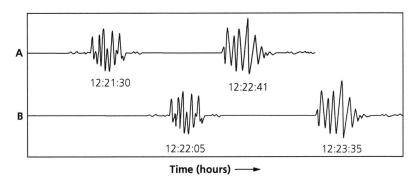

Figure 7.3 Two seismogram traces from different monitoring stations.

Both traces show two signals at different times because the first signal corresponds to the faster P-waves and the second signal is due to the slower S-waves. Station B received the signals after station A, because it is further away from the epicentre of the earthquake. The S–P time lags for each monitoring station are:

$$A_{S–P \text{ time lag}} = 12:22:41 - 12:21:30 = 71\,s$$

$$B_{S–P \text{ time lag}} = 12:23:35 - 12:22:05 = 90\,s$$

A formula can be used to calculate the distance of a monitoring station from the epicentre of the earthquake using the S–P time lag:

$$\text{distance (km)} = \left(\frac{\text{time lag (s)}}{5}\right) \times 60$$

In this case, the distance of the earthquake from each of the stations is:

- distance from A $\text{(km)} = \left(\dfrac{71\,s}{5}\right) \times 60 = 852\,km$

- distance from B $\text{(km)} = \left(\dfrac{90\,s}{5}\right) \times 60 = 1080\,km$

A seismogram from a third seismic monitoring station is required to triangulate the exact epicentre of the earthquake.

> **Exam tip**
>
> Remember to calculate the S–P time (not the other way around). As S-waves arrive second, they arrive at a later time. So, to calculate the time lag, you have to subtract the earlier time from the later time.

Now test yourself

TESTED ☐

7 What is a seismometer?

8 What are the axes on a seismogram?

9 Which type of seismic wave is shown first on a seismogram?

10 What is the S–P time lag on a seismogram?

11 Why are three seismograms, from different monitoring stations, needed to fix the position of an earthquake centre?

12 On 13 October 2016, a 3.1 magnitude earthquake occurred near Merthyr Tydfil. The earthquake occurred at 18 hrs: 09 mins: 12 s. P-waves travelling at 8 km/s propagated away from the epicentre. At what time was the earthquake detected in Wrexham, 145 km away?

Answers on page 120

Seismic waves and the structure of the Earth

Figure 7.4 shows the Earth in cross section.

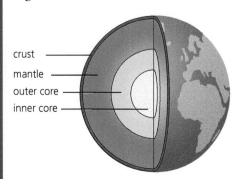

crust
mantle
outer core
inner core

Figure 7.4 The Earth in cross section.

Earthquakes occur in the crust and can travel throughout the Earth. P-waves can propagate through all the layers of the Earth, but S-waves are unable to propagate through the liquid outer core. The speed of P-waves changes with depth inside the Earth as shown in Figure 7.5.

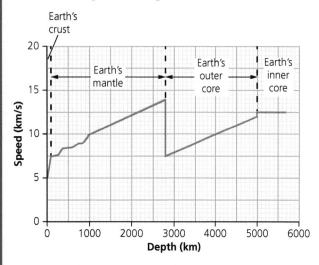

Figure 7.5 The speed of P-waves through the Earth.

The graph in Figure 7.5 shows that the boundary between the solid mantle and the liquid outer core occurs at a depth of 2800 km. The speed of the P-waves increases with depth inside the mantle, due to the increase in pressure and density, causing the seismic waves to refract and bend. As the P-waves cross the mantle–outer core boundary into the liquid outer core, their speed drops rapidly. Figure 7.6 shows the passage of both S-waves and P-waves through the Earth in cross section.

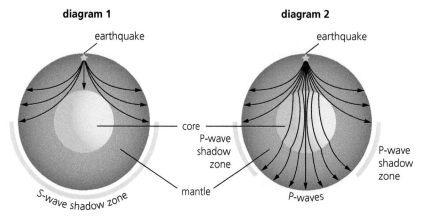

diagram 1   **diagram 2**

earthquake   earthquake

core
P-wave shadow zone
S-wave shadow zone   mantle
P-waves   P-wave shadow zone

Figure 7.6 Seismic waves passing through the Earth.

Exam tip

Both types of seismic waves produce shadow zones. Remember that S-waves cannot propagate through liquids, so do not travel through the outer core, forming a shadow zone directly opposite the Earth from the earthquake epicentre. P-waves produce two shadow zones.

In Figure 7.6, diagram 1 shows S-waves and diagram 2 shows P-waves. As S-waves do not propagate through liquids, there is a large S-wave shadow zone where no S-waves are detected. This is direct evidence for the fact that the outer core is liquid. The P-waves also have shadow zones, due to refraction, particularly as they cross the boundary between the mantle and the outer core.

Now test yourself

TESTED

13 Which seismic waves have a shadow zone on the opposite side of the Earth to an earthquake?
14 Which seismic waves travel through the Earth's mantle layer?
15 Which seismic waves propagate through a liquid?
16 Why are the paths of seismic waves curved?
17 Why does the speed of P-waves change abruptly at a depth of about 2800 km?

Answers on page 120

Summary

- P-waves are longitudinal seismic waves. They are the fastest type of seismic wave and they can travel through solid and liquid rock.
- S-waves are transverse seismic waves. They are slower than P-waves and they can only travel through solid rock.
- Surface waves are seismic waves that form on the Earth's surface. Surface waves are the slowest seismic waves.
- Seismographs are simplified seismic records, which allow for the identification of the lag time

between the arrival of P- and S-waves. Using seismographs from several stations allows the location the epicentre of an earthquake to be determined.
- The path of P-waves and S-waves through the Earth depends on their speed.
- There is an S-wave shadow zone on seismic records. This has led geologists to model the Earth with a solid mantle and a liquid core.

Exam practice

1 When an earthquake occurs, two types of seismic wave, P and S, travel through the Earth.
   (a) State which seismic wave, P or S, cannot travel through a liquid.
   (b) State which seismic wave, P or S, travels faster.
   (c) State which seismic wave, P or S, produces vibrations in the direction of travel of the wave.
   (d) State which seismic wave travels as a longitudinal wave. [4]

WJEC GCSE Physics P3 Foundation Tier Summer 2008 Q1

2 P-waves and S-waves are types of seismic waves produced by earthquakes.
   (a) P-waves and S-waves travel at different speeds.
      (i) Give two other differences between P- and S-waves. [2]
      (ii) Give one difference between transverse and longitudinal waves. [1]
   (b) Figure 7.7 shows signals received from an earthquake at a monitoring station.

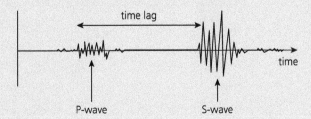

Figure 7.7

      (i) Give a reason why the P-waves are received first. [1]
      (ii) State what information the time lag can tell seismologists about the earthquake. [1]

WJEC GCSE Physics 3 Higher Tier Summer 2009 Q2

3 The map below shows the positions of two seismic recording stations, A and B (in the US state of California). The epicentre of an earthquake lies somewhere on the circumference of the circle around A. Station B is used to locate two possible positions of the epicentre of the earthquake.

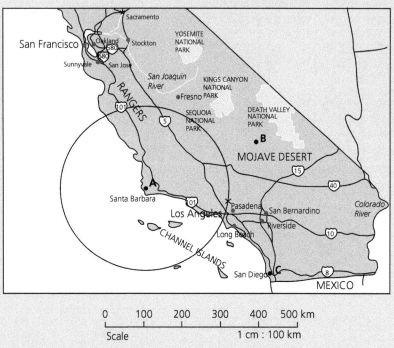

Scale   1 cm : 100 km

Figure 7.8

(a) (i) The P-wave arriving at station B took 25 s to arrive from the epicentre. The speed of the P-wave was 6 km/s. Use the equation below to calculate the distance of B from the epicentre of the earthquake. [2]

distance = speed × time

(ii) One of the possible positions of the epicentre is shown with an X on the circle. Copy the map and mark the other possible position of the epicentre on the circle. [1]

(b) The record of the P-waves and S-waves arriving at station A is shown in Figure 7.9. The S-waves arrived at station A 20 s later than the P-waves.

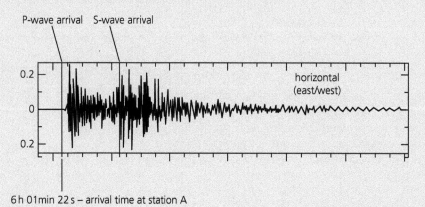

6 h 01 min 22 s – arrival time at station A

Figure 7.9

(i) State why the S-waves arrived later than the P-waves. [1]
(ii) Fill in the gaps below.
The S-wave arrived at station A at _____ hours _____ minutes _____ seconds. [1]

WJEC GCSE Physics P3 Foundation Tier May 2016 Q4

Answers and quick quiz 7 online

ONLINE

8 Kinetic theory

Pressure

REVISED

Pressure is exerted when a force acts over a given area.

$$\text{pressure} = \frac{\text{force}}{\text{area}}$$

$$p = \frac{F}{A}$$

The unit of pressure is the pascal (Pa), where $1\,\text{Pa} = 1\,\text{N/m}^2$. The standard pressure of our atmosphere is also used as a unit of pressure for gases – measured in atmospheres, atm or bar, where $1\,\text{atm}$ or $1\,\text{bar} \approx 1 \times 10^5\,\text{Pa}$. The pressure of a gas is due to the force that the gas particles exert on the walls of its container. The colliding particles exert a force that acts over the area of the walls or the object. Gas pressure is measured with a pressure gauge, such as a Bourdon gauge.

> **Exam tip**
>
> You do not need to remember equations; they are given on the inside cover of the examination paper. You do need to be able to rearrange the equations if you are doing the Higher Tier paper. The Higher Tier equations are shown on page viii.

Now test yourself

TESTED

1 A dog with a weight of 50 N stands on the ground. The total area of the dog's feet is $16\,\text{cm}^2$.
   (a) What is the pressure between the dog's feet and the ground?
   (b) The dog now rears up on its back legs. What is the pressure between the dog's back feet and the ground?
2 Atmospheric pressure at sea level is $1 \times 10^5\,\text{Pa}$. Calculate the force exerted on a wall of a bathroom $2\,\text{m} \times 2\,\text{m}$ by the particles of air inside the room.
3 A football is pumped to a pressure of 1.2 atm, ($1\,\text{atm} = 1 \times 10^5\,\text{Pa}$). The total force of the air particles inside the ball acting on the inside surface of the ball is 15 kN ($1\,\text{kN} = 1000\,\text{N}$). Calculate the inside area of the ball.

Answers on page 120

> **Exam tip**
>
> You do not need to remember unit multiplier prefixes, such as k, kilo; they are printed on the inside of the examination paper. The prefixes are, however, given in standard form, so you do need to know how to use these numbers.

The behaviour of gases

REVISED

A fixed mass of gas expands or contracts when the temperature of the gas alters, changing its volume. Increasing the temperature of the gas speeds up the gas particles and increases the frequency of the collisions of the particles with the walls of its container. This increases the force acting on the walls, increasing the pressure and, if the container is flexible, it expands and the volume increases.

The gas laws

Boyle's law – the relationship between pressure and volume

Experiments show that for a fixed mass of gas at constant temperature, T, pressure, p, is inversely proportional to volume, V. In other words:

$$\text{pressure} \propto \frac{1}{\text{volume}}$$

or

$$\text{pressure} \times \text{volume} = \text{constant}$$

$$pV = \text{constant}$$

This is called Boyle's law and it is illustrated by a graph of pressure against $\dfrac{1}{\text{volume}}$ with a positive correlation.

⊕ Pressure, volume and temperature

Similar experiments to the Boyle's law experiment can be carried out, exploring the link between volume, V, and temperature, T; and pressure, p, and temperature, T.

These experiments show that:

$$\frac{pV}{T} = \text{constant}$$

or

$$\frac{p_1 \times V_1}{T_1} = \frac{p_2 \times V_2}{T_2}$$

where p_1, V_1 and T_1 represent the pressure, volume and temperature of a fixed quantity of gas before a change and p_2, V_2 and T_2 represent the pressure, volume and temperature of the gas after the change. (The temperatures need to be measured as absolute temperatures in kelvin, K.)

> **Exam tip**
>
> Volumes can be given using a range of different units. The standard unit is m³, but this is quite a big volume, so cm³ is used for smaller volumes. Liquids are sometimes given in litres, where 1 litre = 1000 cm³.

Now test yourself

4 A column of air has a volume of 45 cm³ at atmospheric pressure (1×10^5 Pa) and room temperature (293 K). If the temperature remains constant and the volume is compressed to 20 cm³, what is the pressure of the air?

5 The volume of a swimmer's lungs at the surface is approximately 5 litres (1 litre = 1000 cm³) where the pressure is 1 atm (1×10^5 Pa). Every 10 m depth of water increases the pressure on the diver's lungs by 1 atm, so the pressure at 10 m is 2 atm and 20 m is 3 atm, etc. Calculate the volume of the swimmer's lungs at 15 m – assuming that the temperature stays constant.

6 An airship has a volume of 200 m³ as it is inflated at atmospheric pressure and 290 K (17 °C). Gas burners are then used to heat the air inside the airship to a temperature of 510 K. If the airship is allowed to inflate at atmospheric pressure, what is the volume of the airship at 510 K?

7 A high-altitude weather balloon has a volume of 8 litres at ground level, where the air temperature is 293 K and the air pressure is 1×10^5 Pa. Calculate the volume of the balloon at a height of 3.8 km where the temperature is 260 K (−13 °C) and the pressure of the atmosphere is 3394 Pa.

Answers on page 121

Absolute zero

Gases consist of particles moving at high speed in random directions. The higher the temperature, the higher the speed. The particles of a gas collide with the walls of the container. As they collide, they exert a force on the container walls. The force acting over the area of the walls creates a pressure and the pressure of a gas decreases as the temperature decreases ($p \propto T$). As the temperature gets lower and lower, so the gas particles move slower and slower, exerting less pressure on the container. Eventually, at −273 °C, all molecular motion stops and the gas does not exert a pressure on its container. This temperature is known as absolute zero.

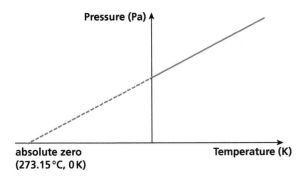

Figure 8.1

> **Exam tip**
>
> The graph showing absolute zero, can also be given as a volume against temperature graph.

Absolute zero is used as the fixed point of the absolute temperature scale, where absolute zero is given the temperature of 0 kelvin (0 K) which is equal to −273 °C.

To convert from kelvin to degrees Celsius:

$$T\,(\text{K}) = \theta\,(°\text{C}) + 273$$

Specific heat capacity

The amount of heat energy required to raise (or lower) the temperature of 1 kg of a material by 1 °C (or 1 K) is called the specific heat capacity of the material, c. Metals have relatively low specific heat capacities (for example, the specific heat capacity of copper is 385 J/kg °C). Non-metals have much higher specific heat capacities (for example, the specific heat capacity of water is 4200 J/kg °C).

The specific heat capacity, c, temperature change, ΔT, the mass, m, and the energy gained (or lost), Q, by a material are related by this equation:

$$Q = mc\Delta T$$

> **Examples**
>
> 1 Calculate the amount of energy required to raise the temperature of 0.75 kg of water inside a kettle from 18 °C to 100 °C.
> 2 A 2.5 kg house brick has a specific capacity of 840 J/kg °C. The brick is heated in an oven adding 85 000 J of heat energy to the brick. Calculate the temperature change of the brick.
>
> Answers
>
> 1 $Q = mc\Delta T = 0.75\,\text{kg} \times 4200\,\text{J/kg}\,°\text{C} \times (100\,°\text{C} - 18\,°\text{C}) = 258\,300\,\text{J}$
>
> 2 $Q = mc\Delta T \Rightarrow \Delta T = \dfrac{Q}{mc} = \dfrac{85\,000\,\text{J}}{2.5\,\text{kg} \times 840\,\text{J/kg}\,°\text{C}} = 40.5\,°\text{C}$

Now test yourself

8 Calculate the amount of energy required to raise the temperature of a room full of air (of mass 61 kg, and specific heat capacity, 1000 J/kg °C) by 22 °C.
9 How much heat energy needs to be removed from 160 g of water in a glass inside a fridge, lowering its temperature from 18 °C to 5 °C?
10 A 0.5 kg block of copper is heated by 70 °C. Calculate the amount of energy needed to do this.

Answers on page 121

Specific latent heat

The amount of energy required to change the state of 1 kg of a material at its melting or boiling point is called its specific latent heat, L. There are two types of specific latent heat: the specific latent heat of fusion (melting or freezing) and the specific latent heat of vaporisation (boiling or condensing). The amount of energy, Q (in J), the mass of the material, m (in kg), and the specific latent heat, L (in J/kg), are related to each other by the equation:

$$Q = mL$$

During a change of state, the energy required to hold the particles together changes. Figure 8.2 shows the relationship between the particles and temperature when solids change to liquids, and then to gases.

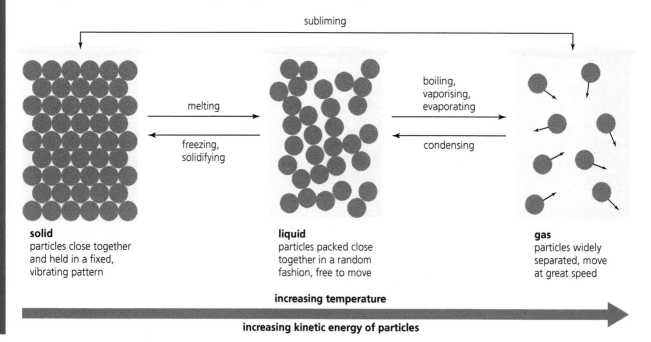

Figure 8.2 Changing state.

When a material changes state, the temperature of the material remains constant. The heat energy added to (or taken away from) the material, rearranges the structure of the particles, as shown in Figure 8.3. The specific latent heat of the material is related to the horizontal sections of the graph, where the material is changing state.

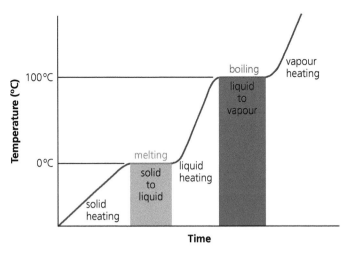

Figure 8.3 Temperature changes during changes of state for water.

Summary

- Pressure is the action of force acting over an area and is given by the equation:

$$\text{pressure} = \frac{\text{force}}{\text{area}}$$

$$p = \frac{F}{A}$$

- A fixed quantity of gas can vary its pressure, volume and temperature. All three quantities are related to each other. Changing one can change the others.
- When a gas is cooled, there is a very low temperature called absolute zero where all molecular motion stops. Absolute zero is used as the zero temperature of the absolute scale of temperature, measured in kelvin, K. A temperature change of 1 K is equal to a temperature change of 1 °C.

 • The pressure, volume and temperature of a gas are related by the equation:

$$\frac{pV}{T} = \text{constant}$$

- The variation of the pressure of gases with volume and temperature can be explained by applying a model of molecular motion and collisions.
- The heat transferred during changes of temperature and state are given by:

$$Q = mc\Delta T$$

$$Q = mL$$

 • A kinetic theory model involving the behaviour of molecules when they are heated can be used to explain the changes in temperature and state of a substance.

Exam practice

1 A fixed mass of gas is kept under conditions of constant volume. The table shows how the pressure of this gas changes with temperature when it is heated.

Temperature (°C)	Temperature (K)	Pressure (N/cm²)
−273	....	0
−173	100	4
−123	150	6
−73	200	8
+27	300	....
+77	350	14
+127	400	16

(a) Copy and complete the table. [2]
(b) Explain in terms of molecules why the pressure increases as the temperature increases. [2]

WJEC GCSE Physics P3 Foundation Tier May 2016 Q5

2 Dan is on holiday in Denver, USA. He packs a sealed plastic water bottle containing only air in his luggage. When he arrives home in Cardiff he notices that the water bottle appears crushed. He works

out the volume of the bottle in both Denver and Cardiff. The table below shows his results together with other relevant information.

The graph in Figure 8.4 shows how atmospheric pressure changes with altitude (height above sea level).

Volume of bottle in Denver	$5.0 \times 10^{-4}\,m^3$
Volume of bottle in Cardiff	$3.9 \times 10^{-4}\,m^3$
Temperature in Cardiff	293 K
Temperature in Denver	293 K
Altitude in Cardiff	0 m

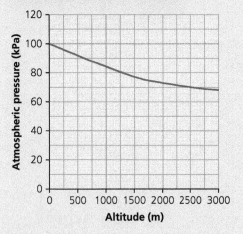

Figure 8.4

(a) (i) Use the graph to write down the air pressure in Cardiff in Pa. [1]

(ii) Use the information above and suitable equations to answer the following questions:

(I) Calculate the atmospheric pressure in Denver and use your answer to find the altitude of Denver. [4]

(II) Calculate the temperature required for the bottle in Cardiff to have the same volume as in Denver. Give your answer in °C. [3]

(b) Explain, in terms of the motion of molecules, how the behaviour of gases leads to the idea of absolute zero and an absolute scale of temperature. [6 QER]

WJEC GCSE Physics P3 Higher Tier May 2016 Q5

3 In Wales about 725 000 plastic bottles are used each day. Plastic bottles that are collected by local councils need to be transported to recycling plants all around Wales. 250 plastic bottles are crushed into a single small bale. This makes it much easier to transfer them to the recycling factory.

A hydraulic press, as shown in Figure 8.5, can be used. It is designed to exert a large force on the plastic bottles to crush them into a compact single bale. Only a relatively small force needs to be applied at X to crush the plastic bottles at Y. The pressure applied on the big piston at Y will be the same as the pressure exerted at X, however the area of the piston at Y is 15 times larger than the area of the piston at X.

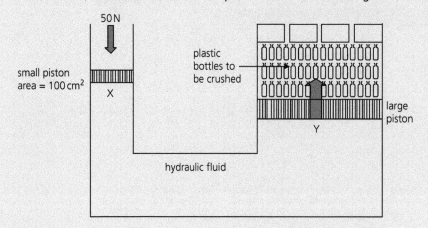

Figure 8.5

➜

(a) If all of the plastic bottles used each day in Wales are crushed, how many small bales would be produced in one week? [2]

(b) Which of the following shows the correct calculation of the pressure exerted by the small piston on the hydraulic fluid at X? [1]

A $pressure = \dfrac{force}{area} = 500 \times 100 = 50\,000 \text{ N/cm}^2$

B $pressure = \dfrac{force}{area} = \dfrac{500}{100} = 5 \text{ N/cm}^2$

C $pressure = \dfrac{force}{area} = \dfrac{500}{100} = 5 \text{ N/m}^2$

D $pressure = \dfrac{force}{area} = \dfrac{100}{500} = 5 \text{ N/cm}^2$

(c) Use information from the text and the equation below to calculate the force applied to crush the plastic bottles at Y. [2]

force = pressure × area

(d) The hydraulic press develops a leak. Hydraulic fluid is expensive. A worker at the recycling factory suggests that replacing the hydraulic fluid with air would save money. Explain why the hydraulic press will no longer work if air is used. [2]

WJEC GCSE Physics Unit 1: Electricity, energy and waves Foundation Tier SAM Q2

4 Metal aerosol cans contain a gas at high pressure. For safety reasons, these cans must be able to withstand pressures up to 620 kPa. A pressure greater than this value will cause the can to explode. A can containing a fixed mass of gas is thrown into a bonfire. It is heated from 27 °C to 227 °C.

(a) Using the model of molecular motion, explain why the pressure of the gas inside the can will increase when thrown into the bonfire. [2]

(b) The original pressure (at 27 °C) of the gas in the can was 280 kPa. Use a suitable equation to determine whether or not the can explodes when thrown into the bonfire. [4]

WJEC GCSE Physics Unit 1: Electricity, energy and waves Higher Tier SAM Q6

5 The information given in the table below states the specific heat capacity of different substances.

Substance	Specific heat capacity (J/kg °C)
Water	4200
Oil	2100
Aluminium	880
Copper	380

(a) (i) Aluminium has a specific heat capacity of 880 J/kg °C. Explain what this statement means. [2]

(ii) A 0.75 kg block of aluminium is heated from 20 °C to 80 °C. Use a suitable equation to calculate the heat energy supplied to the aluminium block. [2]

(b) The hot aluminium block is now submerged into an insulated beaker of water. The mass of the water in the beaker is 0.50 kg. The final temperature of the water and aluminium block is 30.5 °C. Calculate the original temperature of the water. [5]

(c) Explain which of oil or water is a better coolant in a car radiator. [3]

WJEC GCSE Physics Unit 1: Electricity, energy and waves Higher Tier SAM Q7

Answers and quick quiz 8 online

ONLINE

9 Electromagnetism

Magnetic fields

Magnetic fields are places where magnets 'feel' a force. Magnetic fields are indicated using magnetic field lines, which show the pattern of the magnetic field. Magnetic field lines point from North poles to South poles and the closer the magnetic field lines are together, the stronger the magnetic field. Magnetic fields are produced by **permanent magnets**, or they can be produced by electric current flowing through a wire, a coil or a **solenoid**.

Magnetic materials experience a force within **magnetic fields**.

A **permanent magnet** produces its own magnetic field.

A **solenoid** is a long coil of wire.

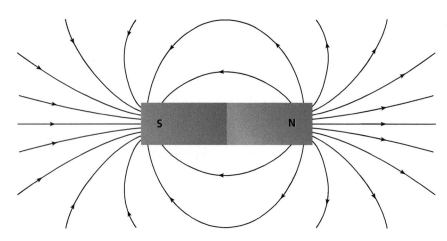

Figure 9.1 The magnetic field around a bar magnet.

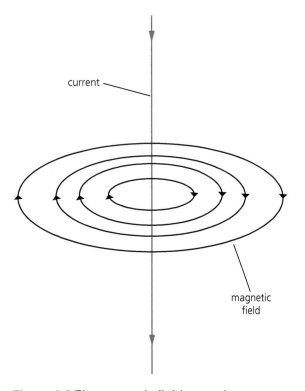

Figure 9.2 The magnetic field around a current-carrying wire.

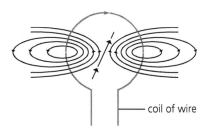

Figure 9.3 The magnetic field around a current-carrying coil.

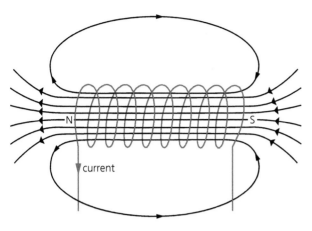

Figure 9.4 The magnetic field around a current-carrying solenoid.

Now test yourself

TESTED

1 Draw the magnetic fields around a bar magnet.
2 How does the magnetic field around a current-carrying wire change as the current increases?
3 List three things that you can do with the magnetic field of a solenoid, that you cannot do with a permanent bar magnet.

Answers on p. 121

The motor effect

REVISED

When a current passes through a wire inside a magnetic field, the wire experiences a force that can move the wire – this is called the motor effect. The direction of the force on the wire depends on the direction of the current and the direction of the magnetic field – and can be determined by using the fingers on your left hand, sometimes called Fleming's left-hand rule.

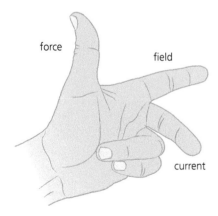

Figure 9.5 Fleming's left-hand rule.

The motor effect is used to design d.c. motors and Fleming's left-hand rule can be used to determine the direction of rotation of the motor. The speed of rotation and the turning force of a d.c. motor increases with: the size of the current; the strength of the magnetic field and the number of turns of wire.

The strength of a magnetic field, B, is related to the density of the magnetic field lines and is measured in tesla, T. A laboratory bar magnet has a magnetic field strength of about 0.1 T. The equation that links the force (F) on a conductor to the strength of the magnetic field (B), the current (I) and the length of conductor (l) is:

$$F = BIl$$

Electric motors

Electric motors convert electrical energy into kinetic energy and they work due to the motor effect. They are designed to convert the force generated into a rotational movement that can drive wheels. Fleming's left-hand rule can be used to determine the direction of rotation of the motor, because it always rotates in the direction of the force.

Now test yourself

4 What does the thumb represent in Fleming's left-hand rule?
5 Calculate the force on a 0.08 m wire, carrying a current of 0.3 A inside a magnetic field strength of 0.7 T.
6 A 7 cm wire experiences a force of 0.5 N inside a magnetic field of 0.3 T. Calculate the current through the wire.
7 List three factors that can increase the speed of a d.c. motor.

Answers on p. 121

Electromagnetic induction

When a conducting wire moves inside a magnetic field, or a magnetic field changes around a conducting wire, an electric current is induced inside the wire – this is called **electromagnetic induction**. The size of the induced current depends on the rate at which the wire cuts the magnetic field lines. Simple a.c. electric generators work due to electromagnetic induction. The electrical output of the generator (current or voltage) increases with: the speed of rotation of the generator coil; the number of turns on the coil and the strength of the magnetic field. The direction of the induced current in a generator depends on the direction of the magnetic field and the direction of rotation of the coil. This can be determined by using the fingers on your right hand (sometimes called Fleming's right-hand rule), where the thuMb points in the direction of the Motion; the First finger points in the direction of the Field and the seCond finger points in the direction of the Current (positive to negative).

> **Electromagnetic induction** is the current induced in a wire by a change in a magnetic field or by the movement of the wire in a magnetic field.

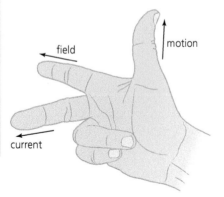

field

motion

current

Figure 9.6 Fleming's right-hand rule.

> **Exam tip**
>
> Do not get confused with Flemings left-hand and right-hand rules. You must learn these by heart. This might help you remember which is which: rIght = Induction; Left = Motor effect (think alphabetical order l for Left, m for Motor, n, o, p...)!

The simple a.c. generator

Knowledge of the process of electromagnetic induction is used to design simple a.c. generators. Generators convert kinetic energy into electrical energy. Inside the generator, a coil of wire rotates inside a magnetic field, or a magnet rotates inside a coil of wire. Electromagnetic induction then induces a voltage in the coil.

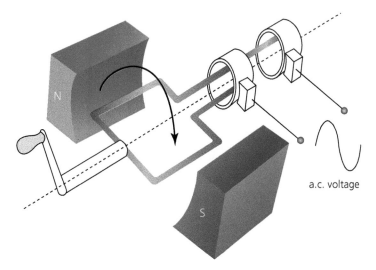

a.c. voltage

Figure 9.7 A simple a.c. generator.

As the coil rotates, the wire cuts across magnetic field lines between the poles of the magnet. This induces a voltage in the wire and a current flows around the coil. Two 'slip' rings connect to the external circuit and the current flows out of the generator. The size of the induced voltage depends on:

- the rate of rotation of the coil
- the number of turns in the coil
- the strength of the magnetic field.

The direction of the induced current depends on the direction of rotation of the coil and the direction of the magnetic field. One side of the wire coil moves up for half a rotation, and then down for the other half of the rotation. This means that for half of a rotation the current flows in one direction, and then flows in the opposite direction for the other half of the rotation. Fleming's right-hand rule allows the actual direction to be determined.

Now test yourself

TESTED

8 What is the effect on the induced current produced by an a.c. generator of the following changes:
   (a) spinning the generator coils faster
   (b) decreasing the strength of the magnetic field
   (c) increasing the number of coils?
9 Figure 9.8 shows a simple a.c. generator. State the direction of the induced current in the wire marked A to B in the coil.

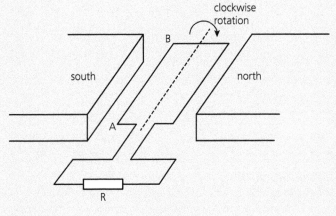

Figure 9.8

Answers on page 123

Transformers

Electric current flowing through a wire causes the wire to heat up. The bigger the current, the greater the heating effect and the greater the energy lost from the wire. The National Grid is a series of wires that allows electrical energy to flow around the country, connecting power stations to users. If the current in the National Grid wires was too high, then the National Grid would be very inefficient in terms of heat loss. To overcome this problem, electricity is transmitted around the country by the National Grid at high voltage, but low current. This is possible because the electricity is transmitted as alternating current, which can be transformed using electromagnetic induction. A typical step-up transformer is shown in Figure 9.9.

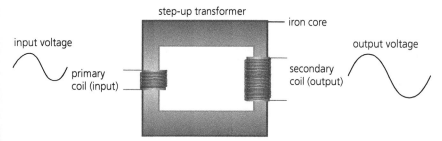

Figure 9.9 A step-up transformer.

An alternating current in the primary coil produces a changing magnetic field in the primary coil. This changing magnetic field is kept within the iron core of the transformer, linking with the secondary coil. The changing magnetic field within the secondary coil induces a voltage in the secondary coil, which generates an alternating current in the secondary coil. The output voltage of the transformer depends on the number of turns on the coils. For an ideal transformer (that is 100 per cent) efficient), the ratio of the voltages and the number of turns is given by:

$$\frac{V_1}{V_2} = \frac{N_1}{N_2}$$

Step-up transformers change low voltage/high current to high voltage/low current and step-down transformers do the reverse.

Example

An ideal (100 per cent efficient) step-up transformer has 80 turns on the primary coil, and 360 turns on the secondary coil. The transformer is fed by an a.c. power supply with a voltage of 12 V. Calculate the voltage on the secondary coil.

Answer

$$\frac{V_1}{V_2} = \frac{N_1}{N_2} \Rightarrow V_2 = \frac{V_1 \times N_2}{N_1} = \frac{12 \times 360}{80} = 54 \text{ V}$$

Now test yourself

10 What does a step-up transformer do to voltage and current?
11 A step-down transformer is designed to convert mains 220 V a.c. to 7 V a.c to charge a tablet. The secondary coil of the transformer has 120 turns. Calculate the number of turns on the primary coil.
12 Why is electricity transmitted as alternating current around the country via the National Grid?

Answers on page 121

Summary

- Bar magnets, straight wires and solenoids all produce characteristic magnetic field-line patterns.
- The strength of a magnetic field is measured in tesla, T, and is determined by the density of the magnetic field lines.
- A magnet and a current-carrying conductor can exert a force on one another (called the motor effect) and Fleming's left-hand rule can be used to predict the direction of one of the following: the force on the conductor, the current and the magnetic field, when the other two are provided.
- **H** The equation that links the force (F) on a conductor to the strength of the field (B), the current (I) and the length of conductor (l), when the field and current are at right angles is:

  $F = BIl$

- The direction of rotation of a simple d.c. motor can be predicted by using Fleming's left-hand rule.
- Increasing the current, magnetic field strength or number of turns on the motor, increases the speed of the motor.

- A current is induced in circuits by changes in magnetic fields and/or the movement of wires. This effect is called electromagnetic induction.
- Electromagnetic induction can be used to explain the operation of a simple a.c. electric generator. Changing the speed of rotation of the generator, the size of the magnetic field and the number of turns on the coil changes the output voltage of the transformer.
- The direction of the induced current in a generator is determined by the direction of the magnetic field and the direction of rotation of the coil. These factors are linked by Fleming's right-hand rule.
- Transformers work due to electromagnetic induction by alternating magnetic fields.
- The output of an ideal (100 per cent efficient) transformer depends on the number of turns on the coils and they are linked by:

  $$\frac{V_1}{V_2} = \frac{N_1}{N_2}$$

Exam practice

1 Figure 9.10 shows a wire being moved between the poles of a magnet.

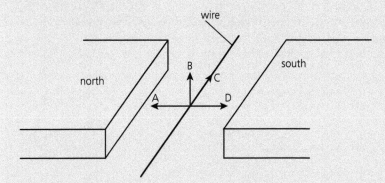

Figure 9.10

(a) Use the letters A, B, C or D to complete the statements that follow.
  (i) The direction of the magnetic field is shown by letter ..... . [1]
  (ii) To produce a current, the wire needs to be moved in direction ..... . [1]
(b) State two ways in which the current in the wire could be made bigger. [2]

WJEC GCSE Physics P3 Foundation Tier Summer 2010 Q2

→

2 Figure 9.11 represents a simple electric motor that a student investigates in his lesson. The current in the coil flows from W to Z as shown.

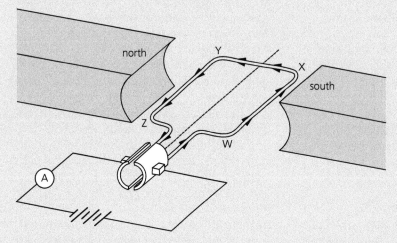

Figure 9.11

(a) (i) Explain clearly how you would use Fleming's left-hand rule to determine the direction of the force on the side YZ. [3]

(ii) State one change the student could make so that side YZ of the coil moves in the opposite direction. [1]

(b) State two changes that could be made to make the coil rotate faster. [2]

WJEC GCSE Physics Unit 1: Electricity, energy and waves Higher Tier SAM Q2

> **Exam tip**
>
> In questions like Question 3 (a), you have to link three boxes on one side with three boxes on the other side. Make sure that you don't draw more than one line from any of the boxes on the left – you will automatically lose one mark.

3 (a) The diagrams on the left in Figure 9.12 show currents flowing in wires of different shapes. The diagrams on the right show the shapes of the magnetic fields produced by currents in wires. Draw lines from the diagrams on the left to the correct field shapes on the right. [2]

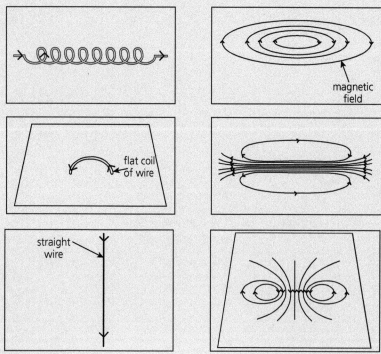

Figure 9.12

→

(b) Use the words from the list below to label a copy of the diagram of a simple d.c. motor shown in Figure 9.13. [3]

carbon brush   magnet   split ring   coil of wire

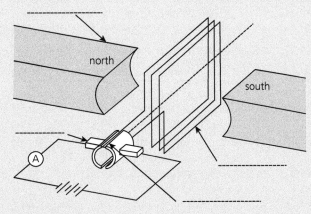

Figure 9.13

(c) Draw an arrow on your diagram to show the direction of the magnetic field (label this arrow as D). [1]
(d) State two ways of making the coil move more slowly. [2]
(e) State one way of reversing the direction in which the coil rotates. [1]

WJEC GCSE Physics P3 Foundation Tier May 2016 Q1

4 (a) State how the construction of a step-up transformer is different from a step-down transformer. [1]

(b) Figure 9.14 shows a transformer that can be used for an investigation in a laboratory. Which of the statements A to E would cause the output voltage to increase? [2]

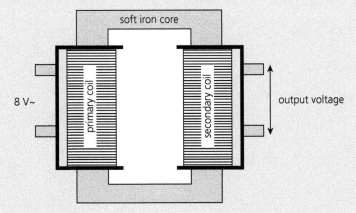

Figure 9.14

A   increasing the number of turns on the primary coil
B   decreasing the number of turns on the primary coil
C   decreasing the input voltage
D   increasing the number of turns on the secondary coil
E   decreasing the number of turns on the secondary coil

(c) Explain why there must be an alternating input voltage for the transformer to work. [2]

(d) An investigation is carried out to determine how the output voltage depends on the number of turns on the secondary coil. The input voltage (8 V) and the number of turns on the primary coil (200) are kept constant throughout the investigation. The results of the investigation are recorded in the table below.

Input voltage (V)	Primary turns	Secondary turns	Output voltage (V)
8	200	50	2
8	200	....	4
8	200	150	6
8	200	200	8
8	200	300	12

(i) Copy and complete the table. [1]
(ii) Plot a graph of the output voltage against the number of secondary turns and draw a suitable line. [3]
(iii) Describe the relationship between the output voltage and the number of secondary turns. [2]
(iv) Use the graph to find the number of secondary turns required to give an output voltage of 5V. [1]
(v) Explain how the graph would be different if the investigation were repeated with a primary coil containing 400 turns. [2]

WJEC GCSE Physics P3 Higher Tier May 2016 Q1

Answers and quick quiz 9 online

ONLINE

10 Distance, speed and acceleration

Describing motion

The motion of an object can be described using the following quantities:

- distance (measured in metres, m): how far the object travels, or how far away the object is from a fixed point
- time (measured in seconds, s): the time interval between two events or the time since the start of the motion
- speed (measured in metres per second, m/s): a measure of how fast or slow the object is moving. The speed of the object can be calculated using the equation:

$$\text{speed} = \frac{\text{distance}}{\text{time}}$$

- velocity (measured in metres per second, m/s, in a given direction): a measure of how fast or slow the object is moving in a given direction (e.g. left/right, North/South), which is the speed in a given direction.
- **acceleration** or deceleration (measured in metres per second per second, m/s^2): the rate that the object is speeding up or slowing down, which is the rate of change of velocity. Acceleration can be calculated using the equation:

$$\text{acceleration or deceleration} = \frac{\text{change in velocity}}{\text{time}}$$

- Speed is a scalar quantity because it only has magnitude (size); velocity is a vector quantity because it has direction as well as magnitude.

> **Acceleration** is the rate of change of velocity.

Now test yourself

1 Calculate the speed of a horse that gallops 200 m in 16 s.
2 Calculate the acceleration of the horse if it takes 5 s to get from rest (0 m/s) to galloping at 12.5 m/s.

Answers on page 121

Graphs of motion

The motion of objects can be described and analysed using graphs of motion. There are two types of motion graph: distance–time graphs and velocity–time graphs.

Distance–time graphs

- A distance–time graph allows us to measure the speed of a moving object.
- The graph in Figure 10.1 shows an object moving away from a starting point at a constant speed of 6 m/s.
- Stationary objects are represented by straight horizontal lines.
- The slope or gradient of a distance–time graph is the speed of the object.

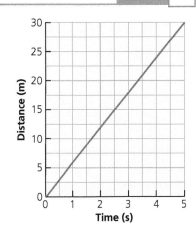

Figure 10.1 Distance-time graph.

Velocity–time graphs

- A velocity–time graph gives us more information than a distance–time graph. The graph in Figure 10.2 shows an object that is:
  - stationary for 2 seconds
  - accelerating at $3\,\text{m/s}^2$ for 2 seconds
  - moving at a constant velocity of $6\,\text{m/s}$ for $6\,\text{s}$.
- The slope or gradient of a velocity–time graph is the acceleration of the object.
- The distance travelled by the object is the area under the velocity–time graph (in this case 42 m).

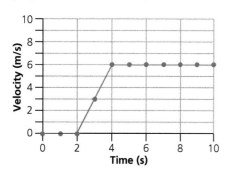

Figure 10.2 Velocity–time graph.

Exam tip

You need to be very careful when you are analysing and extracting information from a graph in a question. Many students make mistakes by reading the axes incorrectly. A good technique to help you measure quantities from graphs is to use a sharp pencil and a ruler to draw thin guidelines onto the axes at the places where you need to take a reading.

Now test yourself

TESTED

3 Sketch distance–time graphs for the following motions:
   (a) an object moving at a constant speed of 6 m/s for 5 seconds
   (b) an object stationary, 20 m away from an observer, for 5 s
   (c) an object moving at a constant speed of 10 m/s for 20 seconds, then 5 m/s for 20 s.
4 Sketch velocity–time graphs of the following motions:
   (a) an object stationary for 2 s, then accelerating at $3\,\text{m/s}^2$ for 2 s, then travelling at a constant speed of 6 m/s for 6 s
   (b) an object initially travelling at 9 m/s and decelerating at $3\,\text{m/s}^2$ for 3 s, then accelerating at $2\,\text{m/s}^2$ for 4 s, then travelling at a constant velocity of 8 m/s for 3 s
   (c) an object initially stationary, accelerates at $3\,\text{m/s}^2$ for 3 s, then travels at a constant velocity of 9 m/s for 4 s, then decelerates $3\,\text{m/s}^2$ for 3 s back to stationary
   (d) an object initially travelling at 6 m/s for 3 s, then accelerates at $2\,\text{m/s}^2$ for 1 s, then remains at 8 m/s for 1 s before decelerating at $2\,\text{m/s}^2$ for 4 s to stationary.

Answers on pages 121–122

Stopping distances

REVISED

Vehicles do not stop instantaneously – there is a time delay between the driver seeing the need to stop, such as a potential hazard, and the vehicle stopping. During this time, the vehicle is still travelling at speed, so it travels through a distance. The total stopping distance of a vehicle is made up of the 'thinking distance' and the 'braking distance'.

- Thinking distance is the distance the vehicle travels while the driver sees the hazard, thinks about braking and then actually reacts by braking.
- Braking distance is the distance that the vehicle moves while the brakes are being applied and the vehicle is decelerating to $0\,\text{m/s}$.

total stopping distance = thinking distance + braking distance

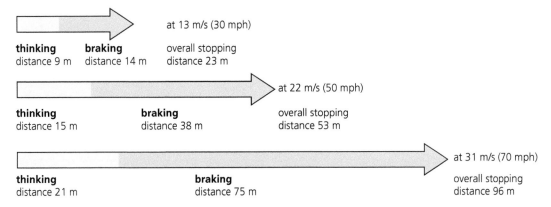

Figure 10.3 Thinking and braking distance at different speeds (from the Highway Code).

Factors affecting stopping distances

Thinking distance depends on several different factors, including:
- the velocity of the car
- the reaction time of the driver (which depends on tiredness, alcohol use, etc.)
- whether the driver is distracted, such as by hearing a mobile phone ring.

The braking distance also depends on several factors:
- the velocity of the car
- the mass of the car
- the condition of the brakes
- the condition of the tyres
- the condition of the road surface
- the weather.

Stopping safely

When a car stops very quickly, for example in a collision, in order to minimise injuries to the occupants, the key factor is to reduce the forces that act on them. Car manufacturers have built safety systems into modern cars to reduce these forces: seat belts, air bags and crumple zones.

Newton's second law (page 72) is:

$$\text{resultant force, } F \text{ (N)} = \frac{\text{change in momentum, } \Delta p \text{ (kg m/s)}}{\text{time for change, } t \text{ (s)}}$$

$$F = \frac{\Delta p}{t}$$

There are two ways of reducing the force on the occupants:
1 by reducing the speed of the collision and so reducing the change of momentum
2 by increasing the time for the collision.

All three safety systems listed above work by increasing the time of the collision – by allowing something to be deformed during the collision. Seat belts stretch; airbags slowly deflate; crumple zones crumple in on themselves.

Now test yourself

5 What is the difference between thinking distance and braking distance?
6 List three factors that affect thinking distance.
7 How do seat belts reduce the force of an impact on a driver?

Answers on page 122

Traffic control measures

The speed of traffic along a road can be controlled by imposing speed limits. Urban areas typically have speed limits of 30 mph, while the national speed limit on single carriageway roads is 60 mph and 70 mph on dual carriageways and motorways. To slow down traffic even more, particularly around schools and residential areas, speed bumps can be installed, forcing drivers to slow down as they drive over the bumps.

Summary

- Speed is a measure of how fast an object is moving:

$$speed = \frac{distance}{time}$$

- Speed is a scalar quantity because it only has magnitude (size).
- Velocity is a vector and has direction as well as magnitude.
- Velocity is measured in metres per second, in a specified direction.
- Acceleration is the rate of change of velocity. Objects that are getting faster are accelerating, and objects that are getting slower are decelerating:

$$acceleration\ or\ deceleration = \frac{change\ in\ velocity}{time}$$

- The units of acceleration are metres per second squared, m/s^2.
- On distance–time graphs stationary objects are shown by flat, straight lines and objects

- travelling at constant velocity are shown by sloping, straight lines.
- The speed of an object can be found by measuring the slope or gradient of the graph.
- For velocity–time graphs, a flat, straight line indicates an object travelling at constant velocity and a straight line sloping upwards indicates an object accelerating. A straight line sloping downwards indicates an object decelerating.
- The acceleration can be found by measuring the gradient or slope of the velocity–time graph.
- The area under the velocity–time graph is the distance travelled.
- The safe stopping distance of a vehicle depends on the driver's reaction time (which affects the thinking distance) and the braking distance of the vehicle.
- Traffic control measures include speed limits and speed bumps.

Exam practice

1 During road tests, three cars are tested to find out how long they take to accelerate from 0 to 60 mph (27 m/s). The results are shown in the table.

Car	Time to reach 60 mph from rest (s)
W	5
X	8
Y	9

(a) State which car, W, X or Y, has the smallest acceleration. [1]
(b) A velocity of 60 mph is the same as a velocity of 27 m/s. Select a suitable equation and use it to calculate the acceleration of car Y during the test in m/s^2. [3]

WJEC GCSE Physics P3 Foundation Tier Summer 2009 Q2

→

2 A theme park ride involves a group of people being lifted in a carriage and then dropped from a height. The graph in Figure 10.4 shows the motion of such a ride.

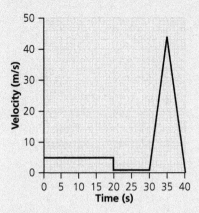

Figure 10.4

(a) Describe the motion of the carriage in the first 20 s. [1]
(b) Select a suitable equation and use it to find the acceleration of the carriage between 30 s and 35 s. [2]

WJEC GCSE Physics P2 Higher Tier Summer 2009 Q3

3 A car overtakes a lorry. In doing so, the car accelerates and, after overtaking safely, returns to its original speed. The graph in Figure 10.5 represents the motion of the car when overtaking the lorry.

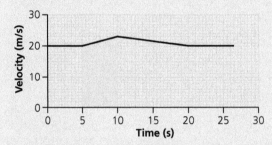

Figure 10.5

(a) Select a suitable equation and then use it, together with data from the graph, to calculate the acceleration of the car during overtaking. [4]
(b) Describe clearly what the shaded area of the graph represents. [2]
(c) Use the data from the graph to calculate the distance travelled between 10 s and 20 s. [3]

WJEC GCSE Physics P3 Higher Tier Summer 2008 Q2

4 The overall stopping distance of a car is made up of two parts: thinking distance and braking distance. At a speed of 20 m/s the Highway Code states that a car has a thinking distance of 12 m and a braking distance of 40 m.

(a) Use a suitable equation to find the thinking time for a driver. [2]
(b) Complete the table below. Some boxes have been completed for you. [3]

Condition	Effect on thinking distance	Effect on braking distance	Effect on overall stopping distance
Poor brakes	No change	Increases	Increases
Driver under the influence of alcohol			Increases
Driver drives at a lower speed	Decreases		
Wet road		Increases	

WJEC GCSE Physics P2 Foundation Tier Summer 2010 Q5

Answers and quick quiz 10 online

ONLINE

11 Newton's laws

Inertia and Newton's first law of motion REVISED

The mass of an object dictates how easy (or difficult) it is to get the object moving or to change its motion. This property is called **inertia**, defined as the resistance of any object to a change in its state of motion or rest. Very massive objects, such as the International Space Station, have very large inertias. Altering their motion takes a very large force.

> **Inertia** is the opposition of an object to a change in its motion.

Newton's first law

In 1687 Isaac Newton realised the link between the motion of an object and the force on it. He summarised this in his first law of motion. On Earth it is very difficult to observe Newton's first law, because friction always acts to oppose the motion of an object.

'An object at rest stays at rest, or an object in motion stays in motion with the same speed and in the same direction, unless acted on by an unbalanced force.'

Resultant forces and Newton's second law of motion REVISED

When several forces act on an object at the same time, they either cancel each other out (balanced forces), or they combine together to produce a resultant (unbalanced) force. Figure 11.1 shows two forces acting on a lorry. The two forces combine to produce a single resultant force of 800 N in the direction of motion.

- A resultant force acting on an object causes a change in its motion. (The lorry will accelerate.)
- Balanced forces cause an object to remain stationary or move at constant speed.

Experiments show us that, as the resultant force on an object increases, so does the size of the acceleration. If we double the resultant force, then the acceleration doubles. The resultant force and the acceleration are proportional to each other. When a graph of resultant force against acceleration is analysed further, we find that the gradient of the line is equal to the mass of the object.

Expressing this as a word equation, we can say:

resultant force, F (N) = mass, m (kg) × acceleration, a (m/s²)

$$F = ma$$

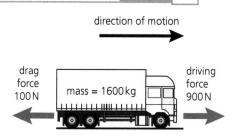

Figure 11.1 Unbalanced forces.

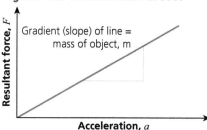

Figure 11.2 A graph of a resultant force against acceleration is a straight line.

Now test yourself   TESTED

1 A lorry, of mass 1600 kg, starts at rest and accelerates to 20 m/s in 40 s. Calculate its:
   (a) acceleration
   (b) resultant force.
2 State:
   (a) Newton's first law of motion
   (b) Newton's second law of motion.

Answers on page 122

Falling objects

The mass of an object is a measure of how much matter (stuff) there is in the object. Mass, m, is measured in kilograms, kg. The weight of an object is the force of gravity acting on the object's mass. Weight is measured in newtons, N. On the surface of the Earth the weight of a 1 kg object is approximately 10 N. The weight of any object on the Earth's surface can be calculated by multiplying its mass, in kg, by 10. The weight of a 1600 kg lorry is therefore 16 000 N.

weight (N) = mass (kg) × gravitational field strength (N/kg)

As an object, such as a parachutist, falls, the weight remains constant. Initially, the only (resultant) force on the parachutist is her weight, so she accelerates downwards. As she speeds up, the force of air resistance acting upwards on her increases. Eventually, it is equal and opposite to her weight – the two forces are equal in size but acting in opposite directions, so they are balanced. The parachutist continues to fall, but at a constant (terminal) speed.

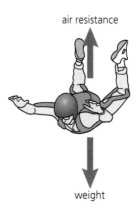

Figure 11.3 Balanced forces mean a moving object travels at constant speed.

Now test yourself

3 Calculate the weight on Earth of an 80 kg skydiver, if g = 10 N/kg.
4 Explain why a skydiver will accelerate when he initially jumps out of an aircraft.
5 Explain why a skydiver will eventually fall at a terminal speed.

Answers on page 122

Interaction pairs

Forces between two objects always act in pairs. The force on one object is called the action force; the force on the other object is called the reaction force. Together they form an interaction pair. The action force and the reaction force are equal and opposite. But they do not cancel each other out because they act on different objects. This is known as Newton's third law:

'For every action force, there is an equal and opposite reaction force.'

For example, in Figure 11.4 some rugby players are practising scrummaging against a static scrum-machine and are pushing with a combined force of 500 N (the action force). Newton's third law means that the scrum-machine exerts a reaction force on the players of 500 N in the opposite direction.

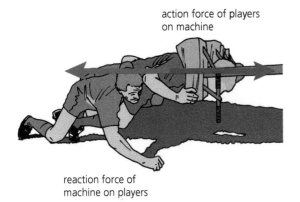

action force of players on machine

reaction force of machine on players

Figure 11.4 The action and reaction forces are equal and opposite.

Some forces (like those involved in scrummaging) are contact forces: the two objects must come into contact with each other to exert the force. Other forces are 'action-at-a-distance' forces, such as gravity, or the forces exerted by electric and magnetic fields. When considering interaction pairs you need to remember that:

1 The two forces in the pair act on different objects.
2 The two forces are equal in size, but act in opposite directions.
3 The two forces are always the same type, for example contact forces or gravitational forces.

Now test yourself

6 State Newton's third law of motion.
7 During a rugby tackle, a tackler exerts a force of 200 N on a tackled player.
  (a) What is the size of the force that the tackled player exerts on the tackler?
  (b) In which direction does the force in (a) act?
  (c) What type of forces are those in this question?

Answers on page 122

Exam tip

When a diagram is given that shows several interaction pairs of forces, it is a good idea to highlight the individual pairs. This makes it easier to analyse them under the pressure of the exam.

Summary

- The mass of an object affects how easy or difficult it is to change the movement of that object. Massive bodies have large amounts of inertia, so require a large force to change their motion, or to make them move if they are stationary.
- Newton's first law of motion states that an object at rest stays at rest or an object in motion stays in motion with the same speed and in the same direction unless acted on by an unbalanced force.
- Newton's second law of motion states that: force = mass × acceleration; that is, the acceleration of an object is directly proportional to the resultant force and inversely proportional to the object's mass.
- Weight is the force of gravity acting on the mass of an object.
- On the surface of Earth, 1 kg of mass has a weight of 10 N; this is called the gravitational field strength, g.

- When an object is falling through the air, initially it accelerates and increases in speed because of the force of gravity acting on it. However, then the force of friction (air resistance) increases, causing the object to decelerate. Eventually, the air resistance force equals the weight of the object and it is said to be falling at its terminal (constant) velocity.
- Newton's third law states: In an interaction between two objects, A and B, the force exerted by body A on body B is equal and oppositely directed to the force exerted by body B on body A.
- Together the action force and the reaction force make an interaction pair.
- Forces may be 'contact' forces, where objects need to come into contact with each other to exert the force, or they may be 'action-at-a-distance' forces, such as gravity or electromagnetic forces.

Exam practice

1 The diagram in Figure 11.3 shows two forces acting on a skydiver. Choose the correct phrase in each set of brackets in the following sentences.
    (a) When the skydiver speeds up, the air resistance is [bigger than / equal to / smaller than] the
        weight. [1]
    (b) When the skydiver falls at the terminal speed, the air resistance is [bigger than / equal to /
        smaller than] the weight. [1]
    (c) When the parachute is opened, the air resistance [gets bigger / stays the same / gets smaller]
        and the skydiver [goes back up / stays in the same place / continues to fall]. [2]

WJEC GCSE Physics P2 Foundation Tier January 2009 Q2

2 Figure 11.5 shows a test rocket on its launch pad. The rocket is powered by three engines, each of which produces a thrust of 2000 N. The mass of the rocket and its fuel is 500 kg, so that its weight is 5000 N.

total thrust from engines

weight of rocket plus fuel

Figure 11.5

    (a) When the engines are fired:
        (i)   calculate the total thrust on the rocket [1]
        (ii) explain why the rocket moves upwards [1]
        (iii) calculate the resultant force on the rocket. [1]
        (iv) Select a suitable equation and use it to calculate the take-off acceleration of the rocket. [3]
    (b) After 2 s, the rocket engines have used up 20 kg of fuel. Assuming that the thrust of the engines
        is constant, calculate:
        (i)   the mass of the rocket and fuel after 2 s [1]
        (ii) the resultant force in newtons on the rocket after 2 s [1]
        (iii) the acceleration of the rocket after 2 s. [1]
    (c) Assuming that the thrust of the engines is constant, explain why the acceleration of the rocket
        will continue to increase for as long as the engines are fired. [2]

WJEC GCSE Physics P2 Higher Tier January 2010 Q4

3 A skydiver of mass 60 kg weighs 600 N.
    (a) The list on the left gives statements about the forces acting on a skydiver falling through the
        air. The list on the right gives five possible effects of these forces on the motion of the skydiver.
        Draw one line from each box on the left to the correct box on the right. [3]

→

The air resistance is greater than the weight.	The skydiver slows down.
The air resistance is equal to the weight.	The skydiver moves upwards.
	The skydiver speeds up.
The weight is greater than the air resistance.	The skydiver falls at constant speed.
	The skydiver stops.

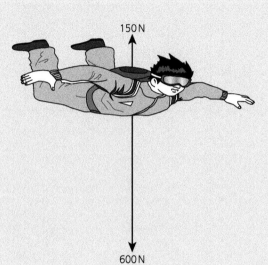

150N

600N

Figure 11.6

(b) Figure 11.6 shows the forces acting on the skydiver at one point in her fall.
  (i) Calculate the resultant force acting on the skydiver. [1]
  (ii) Select and use a suitable equation to calculate the acceleration produced by this resultant force. [2]
(c) The graph in Figure 11.7 shows how the speed changes with time for the skydiver. Choose letters from the graph which complete the following questions.

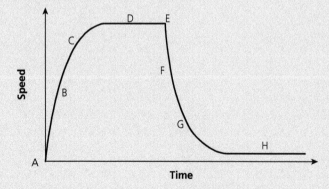

Figure 11.7

(i) The point at which the skydiver opens the parachute is ..... [1]
(ii) The skydiver is at terminal speed with the parachute open at ..... [1]
(iii) Describe and explain the motion of the skydiver in terms of forces. Include in your answer:
  • the type of motion in each region, AB, BC, and CD
  • the forces acting in each region and how they compare. [6 QER]

WJEC GCSE Additional Science/Physics P2 Foundation Tier May 2016 Q1 and Higher Tier Q6(b)

Answers and quick quiz 11 online

ONLINE

12 Work and energy

Work

REVISED

When a force causes an object to move, or acts on a moving object, energy is transferred. The force moves through a distance and energy is transferred as **work**, measured in joules, J.

The amount of work done is calculated by:

work = force × distance moved in the direction of the force

$$W = Fd$$

For example, in rugby, the players lifting the jumper in a lineout exert an upward force, moving the jumper through a distance.

If the lifting force is 1000 N and the player is lifted through 1.5 m then the work done is:

$$W = Fd = 1000 \times 1.5 = 1500\,J$$

The work done is a measure of the energy transferred. But the work done only equals the total energy transferred if no energy is lost as heat to the surroundings (by air resistance or friction).

> **Work** is a measure of the energy transferred.

Figure 12.1 The force, *F*, moves through a distance, *d*.

Gravitational potential energy

REVISED

When an object such as a ball is thrown or kicked vertically, the mass of the ball, *m*, is moved against the Earth's gravitational field strength, *g*, through a change in height, *h*, and it gains gravitational potential energy, PE.

$$\text{gravitational potential energy (PE)} = \text{mass, } m\text{ (kg)} \times \frac{\text{gravitational field}}{\text{strength, } g\text{ (N/kg)}} \times \frac{\text{change in}}{\text{height, } h\text{ (m)}}$$

$$PE = mgh$$

Now test yourself

TESTED

1. Calculate the gravitational potential energy of a 0.44 kg rugby ball kicked vertically upwards to a height of 20 m. The gravitational field strength *g* = 10 N/kg.
2. A rugby player of mass 100 kg is lifted in a lineout, gaining 1500 J of gravitational potential energy. Calculate the height he is lifted. The gravitational field strength *g* = 10 N/kg

Answers on page 122

Kinetic energy

REVISED

When players run with the ball, their muscles convert chemical energy from their food into kinetic energy, KE (movement energy). We can calculate the kinetic energy of any moving object using the equation:

$$\text{kinetic energy (J)} = \frac{1}{2} \times \text{mass, } m\text{ (kg)} \times \text{(velocity, } v)^2\text{ (m/s)}^2$$

$$KE = \frac{1}{2}mv^2$$

> **Exam tip**
>
> The equation for kinetic energy is unusual in that it involves the square of the velocity, v^2. A common mistake is to forget to square the velocity. Square the velocity first, then multiply by the mass and 0.5.

Now test yourself

3 A rugby player can run with a rugby ball with a mean velocity of about 10 m/s. He has a mass of 80 kg. When running at 10 m/s, what is his kinetic energy?

4 A rugby ball of mass 0.44 kg is passed from one player to another with a kinetic energy of 2 J. Calculate the mean velocity of the ball.

Answers on page 122

Total energy

REVISED

When objects such as rugby balls move, there is an interplay of gravitational potential energy and kinetic energy. The total energy of the ball stays constant, assuming that there is no energy lost through air resistance or friction.

total energy = gravitational potential energy + kinetic energy

total energy = PE + KE

Figure 12.2 shows the energy transformations during the flight of a ball after kicking upwards.

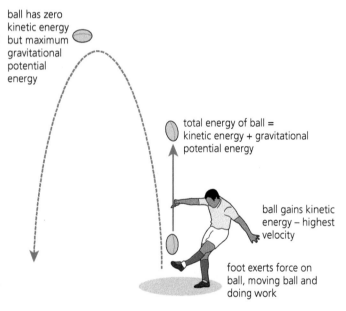

ball has zero kinetic energy but maximum gravitational potential energy

total energy of ball = kinetic energy + gravitational potential energy

ball gains kinetic energy – highest velocity

foot exerts force on ball, moving ball and doing work

Figure 12.2 The interaction of kinetic energy and gravitational potential energy on kicking a ball up.

Exam tip

Questions involving the interaction of gravitational potential energy and kinetic energy often come up in exams. Typically, they involve fairground rides, skiers, or bicycles going down slopes. You should try as many of these types of question as you can, to practise exchanging the two types of energy and doing the calculations.

Storing energy in springs

REVISED

When forces act on springs they can extend (get longer) or compress (get shorter). The spring extension (or compression) depends on the stiffness of the spring (through a value called the spring constant, k) and the force involved. The force, F (in N), spring constant, k (in N/m), and extension (or compression), x (in m), are related to each other using the equation:

force, F = spring constant, k × extension, x

$$F = kx$$

Very stiff springs require a lot of force to extend (or compress) them and they have very high spring constants. A graph of force against extension for a spring, obeying $F = kx$, is shown in Figure 12.3. The gradient (or slope) of the graph line is the spring constant, k, of the spring.

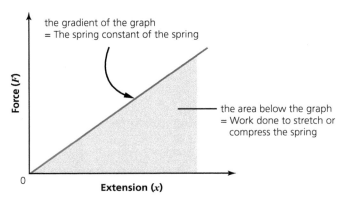

Figure 12.3 **A force–extension graph for a spring.**

When a force is exerted on a spring, it does work on the spring, because work is done when a force moves through a distance (the distance being the extension – or compression – of the spring). The work done stretching (or compressing) the spring can be measured by finding the area under the force–extension graph (a triangular shape). Where the spring obeys the equation $F = kx$, then the work done, W, is given by:

$$W = \frac{1}{2}Fx$$

Now test yourself

TESTED

5 A spring with a spring constant of 25 N/m is extended by 0.14 m. Calculate the force acting on the spring.
6 Explain how a force–extension graph can be used to determine the work done stretching a spring.
7 Calculate the work done stretching the spring in Question 5.

Answers on page 122

Work, energy and vehicles

REVISED

Improving vehicle efficiency

The efficiency of vehicles can be maximised by:
● Improving their aerodynamics, by allowing the air to move more smoothly over the external surfaces of the vehicle. This greatly decreases the amount of work that the vehicle has to do forcing the air away, improving fuel economy and increasing the range of the vehicle.
● Lowering the bottom of the vehicle, making the wheels more enclosed by the wheel arches. This also improves the aerodynamics of the vehicle and reduces the air resistance, improving the fuel economy.
● Designing tyres to ensure a balance between the need for grip (to ensure the safety of the passengers) and the need to minimise the rolling resistance between the tyre and the road surface (also improving the fuel economy).
● Reducing the amount of energy lost when the vehicle is idling in traffic or at traffic lights. Computers monitor the engine and, if it is forced to idle when the vehicle stops, systems operate to temporarily shut down the engine, or the kinetic energy of the idling engine is used to power a small dynamo that charges a battery, which is later used to power the vehicle's electrical systems.

- Using new lightweight materials to replace heavier, traditional materials, such as steel, for vehicle parts. Reducing the mass of the car reduces its inertia to movement and so less energy is needed to get the vehicle moving in the first place.

Improving vehicle safety

One of the factors that affects the amount of damage to drivers and passengers in a car crash is the rapid deceleration that occurs when a car crashes. If you are in a car that suddenly slows down, your motion keeps you moving forward until a force acts to change your velocity (Newton's first law) – this could be the force between your head and the windscreen!

From Newton's second law, where:

force (N) = mass (kg) × acceleration (m/s^2)

if you decelerate very quickly, there will be a large force on your body. Anything that increases the time taken for the collision, and in so doing reduces the deceleration, reduces the force acting on you. The trick is to engineer systems within the car that increase the time for a collision and yet keep the passengers safe inside a strong passenger compartment. Car makers design their cars so that they will collapse gradually on impact (crumple zones) – significantly increasing the time of the collision (reducing the acceleration) and substantially reducing the force on the occupants.

Now test yourself

TESTED

8 List three ways to improve the efficiency of a car.
9 Explain how a crumple zone reduces the impact force on a driver during a head-on collision.

Answers on page 122

Exam tip

A common extended-answer exam question asks you to explain how car safety systems keep a driver and any passengers safe during a collision. Seat belts, crumple zones and air bags are good examples to choose, and they all work by extending the collision time and reducing the forces that act during a collision.

Summary

- When a force acts on a moving body, energy is transferred although the total amount of energy remains constant.
- The equation for work is:

  work = force × distance moved in the direction of the force

  $W = Fd$

- Work is a measure of the energy transfer, so that work = energy transfer (in the absence of thermal transfer).
- An object can possess energy because of its motion (kinetic energy) and position (gravitational potential energy) and deformation (elastic energy).
- The equation for kinetic energy is:

  kinetic energy = $\frac{1}{2}$ × mass × (velocity)^2

  $KE = \frac{1}{2}mv^2$

- The equation for change in potential energy is:

  change in potential energy = mass × gravitational field strength × change in height

  PE = mgh

- The relationship between force and extension for a spring is:

  force = spring constant × extension

  $F = kx$

- The work done in stretching can be determined by finding the area under the force–extension (F–x) graph; $W = \frac{1}{2}Fx$ for a linear relationship between F and x.
- The energy efficiency of vehicles can be improved by reducing aerodynamic losses/air resistance and rolling resistance, idling losses and inertial losses.
- Seat belts, air bags and crumple zones on vehicles all work by extending the time of a collision and reducing the force acting on the occupants of the vehicle.

Exam practice

1 Figure 12.4 shows a low-loader lorry winching a car up a ramp. The winch of the lorry does 2450 J of work in lifting the car and 350 J of work against friction while pulling the car up the 3.5 m ramp.

Figure 12.4

(a) Calculate the total work done in raising the car onto the back of the lorry. [1]

(b) Select and use a suitable equation to find the force, *F*. [3]

WJEC GCSE Physics P2 Foundation Tier Summer 2010 Q7

2 Figure 12.5 shows a winch at Y which is used to pull a yacht at X, 50 m up a slipway, through a vertical height of 4 m.

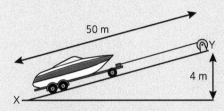

Figure 12.5

(a) The weight of the yacht is 15 000 N, and it is lifted through the vertical height of 4 m. Select and use a suitable equation to calculate the work done against gravity, lifting the yacht through 4 m. [2]

(b) A frictional force of 1000 N acts on the yacht as it is pulled up the 50 m slipway. Use your equation from (a) to calculate the work done against this frictional force. [1]

(c) (i) Hence, calculate the total amount of work done by the winch in pulling the yacht up the slipway. [1]

(ii) Calculate the force that must be applied by the winch in pulling the yacht up the slipway. [2]

WJEC GCSE Physics P2 Higher Tier Summer 2007 Q7

3 A lift takes people up to a jump platform in a bungee tower. The jump platform is 55 m above the ground.

(a) The lift takes a 60 kg person from the ground to the jump platform. Select a suitable equation and use it to find the increase in gravitational potential energy of the person. (Gravitational field strength = 10 N/kg) [3]

(b) The bungee jumper has a kinetic energy of 18 000 J when he is falling at maximum speed.

(i) What is his potential energy when he reaches his maximum speed? [1]

(ii) Select and use an equation to find his maximum speed. [3]

(c) Explain in terms of named forces why the speed increases before the bungee rope starts to stretch. [2]

(d) The bungee rope stretches and stops the jumper just above ground level, storing the bungee jumper's energy in the rope. Give the values at this point of:

(i) his kinetic energy [1]

(ii) his gravitational potential energy [1]

(iii) the energy stored in the bungee rope. [1]

WJEC GCSE Physics P2 Higher Tier January 2008 Q5

Answers and quick quiz 12 online

ONLINE

13 Further motion concepts

Momentum

REVISED

Another way to link force and motion is to think of both the mass and the velocity of a moving object. Objects that require a large force to stop either have very large mass (inertia) or are moving at very high velocity. **Momentum** is the name given to the product of the mass and velocity of an object. So objects with a large momentum require a large force to alter their motion.

> **Momentum** is the product of mass × velocity.

momentum, p (kg m/s) = mass, m (kg) × velocity, v (m/s)

$$p = mv$$

Now test yourself

TESTED

1 A rifle fires a bullet of mass 0.005 g (5×10^{-6} kg) with a velocity of 400 m/s. Calculate the momentum of the bullet.
2 The rifle has a mass of 5 kg and recoils with the same momentum. Calculate the recoil velocity of the rifle.

Answers on page 122

> **Exam tip**
>
> When using an equation to calculate a value such as momentum, you need to be careful about which numbers you use. In the exam, once you have chosen and written down the equation, you could highlight the quantity in the equation and the appropriate number in the question, perhaps using the same coloured highlighter pen.

Newton's second law and momentum

REVISED

Newton's second law, $F = ma$, can be explained and written in a different way using the change in momentum of an object. You will remember that the momentum of an object is equal to the mass of the object multiplied by its velocity. When an object accelerates, it changes its velocity from one value to another. This means that, when it accelerates, it also changes its momentum; the resultant force acting on an accelerating object is equal to the rate of change of momentum of the object.

resultant force, F (N) = $\dfrac{\text{change in momentum, } \Delta p \text{ (kg m/s)}}{\text{time for the change, } t \text{ (s)}}$

$$F = \frac{\Delta p}{t} = \frac{\Delta mv}{t}$$

where

Δp = final momentum − initial momentum = $mv_{final} - mv_{initial}$

and

$$F = \frac{\Delta p}{t} = \frac{mv_{final} - mv_{initial}}{t}$$

Now test yourself

TESTED

3 Calculate the resultant force acting on a skateboarder who changes his momentum from 35 kg m/s to 53 kg m/s in 3 s.
4 A force of 240 N acts on a yacht for 15 s. Calculate the change in momentum of the yacht.

Answers on page 122

The law of conservation of momentum

Momentum is a very important quantity within the Universe. Experiments show that momentum is a quantity that is always conserved whenever objects interact with each other (either by collision or explosion). This applies to interactions between stars and galaxies at one end of the size scale and to sub-atomic particles, like protons and electrons, at the other end of the scale. The law of conservation of momentum can be written as:

total momentum before an interaction = total momentum after an interaction

(By convention, the momentum in one direction is considered to be positive and momentum in the opposite direction is considered to be negative.)

Conserving kinetic energy

You will remember that the kinetic energy of a moving object is given by:

 $KE = \frac{1}{2}mv^2$

where m is the mass of the object (in kilograms) and v is the velocity of the object (in m/s). Although momentum is always conserved in collisions, kinetic energy is only conserved in elastic collisions, where the total kinetic energy before the collision is the same as the total kinetic energy after the collision. However, very few (if any) collisions are truly elastic. Most involve some kinetic energy loss, which is transformed into other forms of energy, such as the strain energy that deforms an object, or heat and sound – these collisions are called inelastic collisions.

> ## Now test yourself
>
>
> 5 State the law of conservation of momentum.
> 6 A lawn-bowls jack ball of mass 0.25 kg, travelling at 3 m/s, hits a larger lawn bowl of mass 0.75 kg and stops dead. Calculate the velocity of the larger lawn bowl following the collision.
>
> Answers on page 122

Equations of motion

The motion of an object can be completely described by four (kinematics) equations, provided that the object is moving with a constant (or zero) acceleration.

The equations use a set of standard symbols for each of the motion quantities:
- x – distance travelled (in m)
- u – initial velocity of the object (in m/s)
- v – final velocity of the object (in m/s)
- a – acceleration of the object (in m/s²)
- t – time of the motion (in s)

> **Exam tip**
>
> You need to learn (memorise) the meaning of each of the symbols that are used in the kinematics equations.

The equations are:

$$v = u + at$$

$$x = \left(\frac{u + v}{2}\right)t$$

$$x = ut + \frac{1}{2}at^2$$

$$v^2 = u^2 + 2ax$$

When objects are moving by falling due to the Earth's gravitational field, the symbol g is normally used instead of a, and the value of g is $10\,\text{m/s}^2$.

Now test yourself

TESTED

7 (a) A girl on a bicycle, travelling at $1.5\,\text{m/s}$, accelerates at $0.5\,\text{m/s}^2$ for $3\,\text{s}$. Calculate the final velocity of the girl on the bicycle.
  (b) Calculate the distance travelled by the girl on the bicycle during the $3\,\text{s}$ interval.

Answers on page 122

Turning forces

REVISED

When a force, F, acts at a distance, d, away from the centre of a pivot (such as a spanner turning a nut), the force exerts a turning effect on the pivot, called a turning force.

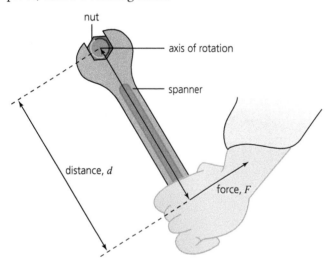

Figure 13.1 **Momentum of a force.**

The turning force that acts on the pivot is called a moment and it is calculated using the equation:

moment, M = force, F × distance, d

$$M = Fd$$

where d is the distance of the force away from the axis of rotation of the pivot. The units of a moment are Nm.

The principle of moments

In many cases, moments tend to act in pairs about a pivot. The classic example of this is a see-saw with two people sitting one at each end:

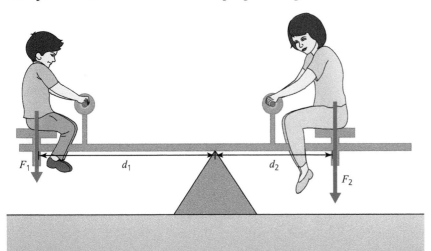

Figure 13.2 Children balancing on a see-saw.

Exam tip

Take care when answering questions involving the principle of moments, as the units may be mixed. For example, distances could be given in cm, rather than m. Make sure you use the same distance unit in the calculation of the clockwise and the anticlockwise moments.

When an object like the see-saw is in balance, the sum of the clockwise moments (in this case $M_{clockwise} = F_2 d_2$) equals the sum of the anticlockwise moments (in this case, $M_{anticlockwise} = F_1 d_1$) about a point; this is called the principle of moments:

sum of clockwise moments, $M_{clockwise}$ = sum of anticlockwise moments, $M_{anticlockwise}$

In the case of the see-saw in Figure 13.2:

$$F_1 d_1 = F_2 d_2$$

Now test yourself

8 State the principle of moments.
9 A decorator uses a screwdriver to open the lid of a can of paint, exerting a force of 8 N, 0.3 m from the edge of the paint tin. Calculate the turning force acting on the lid.
10 A crane lifts a weight of 3000 N, 12 m away from the crane tower. The weight is balanced on the other side of the crane tower by a counter-weight, 4 m from the tower. Calculate the weight of the counter-weight.

Answers on page 122

Summary

- The momentum of a body depends on its mass and its velocity, and is determined by the equation:

  momentum = mass × velocity

  $$p = mv$$

- Newton's second law of motion can be written as:

  resultant force,

  $$F(N) = \frac{\text{change in momentum, } \Delta p \text{ (kg m/s)}}{\text{time for the change, } t \text{ (s)}}$$

  $$F = \frac{\Delta p}{t} = \frac{\Delta mv}{t}$$

- The law of conservation of momentum states that momentum is always conserved during the motion of objects and can be used to perform calculations involving collisions or explosions between objects.
- The kinetic energy equation, $KE = \frac{1}{2}mv^2$ can be used to compare the kinetic energy before and after an interaction.
- The motion of objects can be modelled using the equations:

$$v = u + at$$

$$x = \left(\frac{u + v}{2}\right)t$$

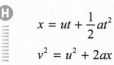

$$x = ut + \frac{1}{2}at^2$$

$$v^2 = u^2 + 2ax$$

- A turning force causes rotation and is called a moment of the force, where moment = force × distance (between the axis of rotation and the force):

$$M = Fd$$

- The principle of moments states that, if a beam is balanced, then the sum of the clockwise moments must equal the sum of the anticlockwise moments about a point.

Exam practice

1 A gun fires a bullet of mass 0.01 kg, with a speed of 1000 m/s, at a target. The bullet passes straight through the target, losing some momentum as it does so, before emerging with a velocity, v.
   (a) (i) Select and use a suitable equation to calculate the momentum of the bullet before passing through the target. [2]
   (ii) The bullet loses 9 kg m/s of momentum passing through the target. Calculate the momentum of the bullet as it emerges from the target. [1]
   (iii) The emerging bullet has the same mass as it did before entering the target. Use your answer to part (ii) to calculate the velocity of the bullet as it emerges from the target. [2]
   (b) The gun has a mass of 1.25 kg. When the bullet is fired, the gun recoils with the same value of momentum as the fired bullet (your answer to part (a) (i)). Use this information to calculate the recoil velocity of the rifle. [2]

   WJEC GCSE Physics P3 Higher Tier Summer 2008 Q6

2 The diagrams in Figure 13.3 show two stationary space vehicles in the act of separating.
   Vehicle A has a mass of 50 000 kg.
   Vehicles A and B are at rest before the separation. The total momentum is zero.
   After the separation, vehicle A moves with a velocity of −2 m/s.

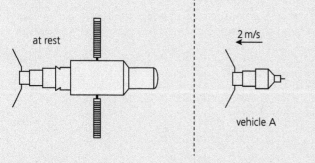

at rest

2 m/s

vehicle A

vehicle B

Figure 13.3

   (a) Use the equation below to calculate the momentum of A after the separation. [2]

   momentum = mass × velocity

   (b) No momentum is lost when they separate. Write down the momentum of B after they separate. [1]

(c) Vehicle B has a mass of 80 000 kg. Use the equation below to find the velocity of vehicle B after the separation. [2]

$$velocity = \frac{momentum}{mass}$$

WJEC GCSE Physics P3 Foundation Tier May 2016 Q3

3 A small stone falls through the air. The table below shows how the velocity of the stone changes in the first 4 seconds.

Time (s)	Speed (m/s)
0	0
1	10
2	20
3	30
4	40

(a) State why the distance travelled between 2 seconds and 4 seconds is bigger than for the first two seconds. [1]

(b) (i) Use information from the table above and the equation below to calculate the acceleration. [2]

$$a = \frac{v - u}{t}$$

(ii) Use information from the table above and the equation below to calculate the distance that the stone fell between 2 s and 4 s. [2]

$$x = \frac{1}{2}(u + v)t$$

(c) If a feather were dropped instead, its velocity after 4 s would be less than that of the stone. Give a reason for the difference. [1]

WJEC GCSE Physics P3 Foundation Tier May 2016 Q6

4 A football of mass 0.3 kg is dropped from rest off a bridge and takes 2.8 seconds to reach the ground below. Select and use suitable equations to answer the questions below. Assume the acceleration due to gravity = 10 m/s² and that air resistance is negligible.

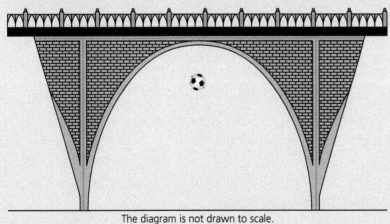

The diagram is not drawn to scale.

Figure 13.4

(a) Calculate the height of the bridge. [2]
(b) Calculate the momentum of the ball just before it hits the ground. [3]
(c) The ball rebounds from the ground with a speed of 14 m/s.
   (i) Calculate the kinetic energy of the football as it leaves the ground. [2]
   (ii) Calculate the change in momentum of the ball due to the bounce. [2]
   (iii) Explain how momentum is conserved when the ball rebounds from the Earth. [2]
(d) Describe how Newton's third law of motion applies when the ball hits the ground. [2]

WJEC GCSE Physics P3 Higher Tier May 2016 Q4

5 See-saws, tower cranes and even simple levers are all things that rely on an understanding of balance and moments in their design. For objects not to topple, the moments must balance. A class of students studying moments carried out the following experiment. They set up a metre rule to balance at its midpoint and then placed weights at different distances from the centre to get it to level. One of the weights was kept the same throughout at 5 N but its distance, d, from the pivot (centre of the rule) could be changed. The other balancing weight, W, could be varied but its distance from the pivot was kept constant at 20 cm. This is shown in Figure 13.5.
The results from one group are shown in the table.

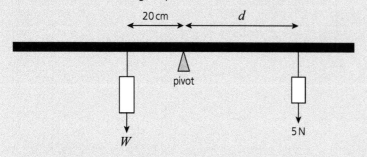

Figure 13.5

Left side		Right side	
W (N)	Distance (cm)	Weight (N)	d (cm)
3	20.0	5	12.0
4	20.0	5	16.0
5	20.0	5	20.0
8	20.0	5	32.0
10	20.0	5	40.0
12	20.0	5	48.0

(a) (i) Complete the moments equation for the situation shown in Figure 13.5. [1]

$$W \times \text{............} = 5 \times d$$

   (ii) Plot a graph of the values of W from the left side against values of d from the right side. [3]
   (iii) Give the value of the weight, W, that would balance at a distance, d, of 10 cm. [1]
   (iv) Give the value of d that would balance a weight, W, of 6 N. [1]
   (v) Describe how the weight, W, changes as the distance, d, changes. [2]
   (vi) Use your graph to explain whether further readings should have been taken in this experiment. [2]

(b) The metre rule shown in Figure 13.6 is supported at its midpoint. A student suggests that it is balanced. Use the principle of moments to investigate this claim. [3]

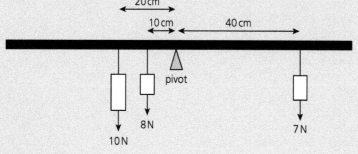

Figure 13.6

WJEC GCSE Physics Unit 2: Forces, space and radioactivity Higher Tier SAM Q3

Answers and quick quiz 13 online

ONLINE

14 Stars and planets

The scale of the Universe

The Solar System

The Universe is a very big place – it would take about 13.75 thousand million (13.75 billion) years for light to travel from the Earth to the edge of the observable Universe. Our local patch of the Universe is called the Solar System. The main constituents of the Solar System are:

- 1 star – the Sun
- 8 planets – Mercury, Venus, Earth, Mars, Jupiter, Saturn, Uranus and Neptune
- 146 moons (a moon is a natural satellite of a planet)
- 5 dwarf planets, including Pluto
- an asteroid belt – between Mars and Jupiter
- many comets and other small lumps of rock and interplanetary dust.

Planets

Of the eight planets, the inner four are rocky planets (Mercury, Venus, Earth and Mars), and the outer four are gas giants (Jupiter, Saturn, Uranus and Neptune). Of the four rocky planets, Venus, Earth and Mars have atmospheres, and the gas giants are mostly composed of hydrogen, helium and some methane. Data on the eight planets is given in the table below.

Planet	Symbol	Mean orbit radius (in AU)	Orbital period (in Earth years)	Mean radius in ($R_\oplus$)	Mass in ($M_\oplus$)
Mercury	☿	0.39	0.24	0.38	0.06
Venus	♀	0.72	0.62	0.95	0.82
Earth	⊕	1.0	1.0	1.0	1.0
Mars	♂	1.5	1.9	0.53	0.11
Jupiter	♃	5.2	12	11	320
Saturn	♄	9.6	29	9.5	95
Uranus	♅	19	84	4.0	15
Neptune	♆	30	170	3.9	17

Now test yourself

1 Explain how the four inner planets are different from the four outer planets.
2 What is the relationship between the orbital radius and orbital period for the planets?

Answers on pages 122–123

> **Exam tip**
>
> You need to learn the order of the planets away from the Sun. My Very Easy Method Just Speeds Up Naming!

Measuring distances in the Universe

- Earth radius, $R_\oplus$ – the size of a planet is measured relative to the Earth, so the radius of Jupiter = $11\,R_\oplus$. Comparisons to the Earth's dimensions are good measurements to use to compare the planets.
- Astronomical Units, AU – this is the average distance of the Earth from the Sun. Distances in the Solar System are measured using this unit. Neptune, the furthest planet, is 30 AU from the Sun and the outer reaches of the Solar System stretch out to over 100 000 AU. ($1\,AU = 1.5 \times 10^{11}\,m$)
- Light years, ly – the light-year is the distance that light travels in one year – $9.47 \times 10^{15}\,m$. This unit is used to measure distances to our nearest stars and within our own galaxy of stars, the Milky Way. The closest star to the Sun, Proxima Centauri, is 4.2 ly away. Our Solar System is about 4 ly in diameter, and the Milky Way galaxy is about 100 000 ly across. Our galaxy is part of a 'Local Group' of galaxies, about 10 million ly across, and the Local Group is part of the Virgo Supercluster of galaxy clusters, about 110 million ly across. The Virgo Supercluster is one of the largest observed structures in the Universe. The edge of the observable Universe is 13 750 million ly away.

> **Exam tip**
>
> You do not need to learn the conversion factors for metres into AU and ly. You would be given these values in the question paper.

Now test yourself

TESTED

3 What is the distance of Jupiter away from the Sun:
  (a) in AU
  (b) in m
  (c) in ly?

Answers on page 123

How did the Solar System form?

The Sun and the Solar System formed out of the nebula (a gas and dust cloud) that resulted from the supernova death of a huge star. As our original nebula collapsed in on itself due to gravity, denser, darker regions appeared, where protostars formed. A protostar is part of a nebula, collapsing due to gravity, and it is the stage in a star's formation before nuclear fusion starts.

As the protostar collapsed further due to gravity, more gas and dust were drawn into it from the surrounding nebula. The pressure inside its core rose enough for the temperature to exceed 15 million °C and nuclear fusion reactions of hydrogen gas started and a star was born.

Stellar lifecycles and the Hertzsprung–Russell diagram

REVISED

The vast majority of stars spend most of their lives as **main-sequence stars**. The term 'main-sequence' was first coined by the Danish astronomer Ejnar Hertzsprung in 1907 who realised that the colour (or spectral class) of a star correlated to its apparent brightness (the brightness of a star as seen from Earth) and that many stars appeared to follow a simple relationship between these two variables. At the same time, an American astronomer called Henry Norris Russell was studying how the spectral class varied with actual (or absolute) brightness by correcting the brightness of stars for their distance away from Earth. The diagram showing absolute stellar brightness (or luminosity) against stellar temperature (which dictates the spectral class or colour of a star) is now known as the **Hertzsprung–Russell (HR) diagram** (Figure 14.1).

> A **main-sequence star** is one that releases energy by fusing hydrogen into helium.
>
> The **Hertzsprung–Russell (HR) diagram** is a means of displaying the properties of stars and depicting their evolutionary paths.

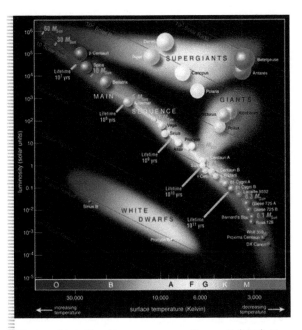

Figure 14.1 The Hertzsprung–Russell (HR) diagram.

The HR diagram arranges stars into groups according to their properties. The axes of the HR diagram in Figure 14.1 are luminosity (or the total amount of light energy emitted by the star, in solar units, where the luminosity of the Sun = 1) on the y-axis, and temperature (in kelvin) displayed as a non-linear power (of ten) series along the x-axis. The x-axis (which, unusually for a graph, runs backwards in temperature from left to right) can also be displayed as colour – red stars being coolest and blue stars being hottest. The HR diagram can be divided into four distinct quadrants, shown in Figure 14.2.

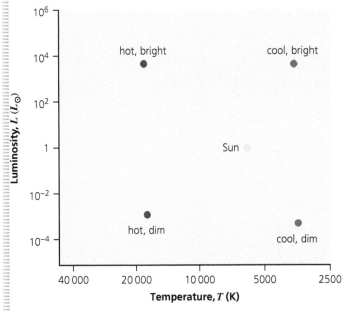

Figure 14.2 The four quadrants of the Hertzsprung–Russell (HR) diagram.

The hottest, brightest stars on the HR diagram are the largest main-sequence stars. The cool, brightest stars are red super-giants. The hot, dim stars are all white-dwarf stars, and the coolest, dim stars are red-dwarf stars.

Now test yourself

4 What are the axes of a HR diagram?
5 Which groups of stars have the following properties:
   (a) cool, dim stars
   (b) hot, dim stars
   (c) hot, bright stars
   (d) cool, bright stars?
6 Where are main-sequence stars found on the HR diagram?

Answers on page 123

Stellar lifecycles

Main-sequence stars run from top left to bottom right on the HR diagram. Above the main-sequence stars are the giant stars, which have a radius of between 10 and 100 times that of the Sun. A **red giant** is a dying star in one of the last stages of its evolution. A main-sequence star swells up to form a red giant when it starts to run out of hydrogen fuel and starts to fuse helium gas. Red giants are an important stage in the lifecycle of most main-sequence stars. Most stars spend most of their lifetime on the main sequence where their stability depends on the balance between the gravitational force of attraction trying to implode the star, and the combination of gas and radiation pressure trying to push the star outwards. Radiation pressure is the effect of the electromagnetic radiation moving out from the core of the star.

When the hydrogen nuclear fuel of the main-sequence star starts to run out, nuclear fusion of helium takes over in the core. The radiation pressure increases and the balance of stability of the star is tipped in favour of the star expanding. As it gets bigger, the energy produced by the star is spread over a much larger surface area; its surface temperature drops; its colour becomes redder and it becomes a red giant. Red-giant stars are unstable, and the star increasingly relies on the nuclear fusion of heavier and heavier elements (a process called nucleosynthesis). Once the fusion reactions have produced the element iron, the star cannot gain energy from forming heavier elements and fusion ceases. The star collapses; its outer atmosphere is puffed outwards as a planetary nebula and the remaining hot core is termed a **white dwarf**. White-dwarf stars are found in the bottom left quadrant of the HR diagram. A white dwarf no longer undergoes nuclear fusion, but gives out light because it is still very hot.

The rest of the star's lifetime is a cooling process, as it is no longer generating energy via nuclear fusion. The white dwarf cools, moving to the right on the HR diagram, forming a red dwarf, and ultimately a black dwarf. The lifecycle plot of the Sun on the HR diagram, showing how its luminosity and surface temperature will change over its lifecycle, is shown in Figure 14.3.

> A **red giant** is a very large star which fuses helium into heavier elements.
>
> A **white dwarf** is a star at the end of its life. No fusion occurs, it is cooling down.

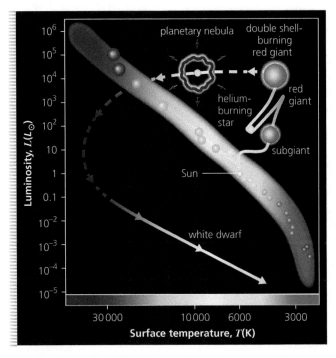

Figure 14.3 The HR diagram lifecycle plot of the Sun.

> **Exam tip**
>
> You need to learn the lifecycle of the Sun: protostar → main sequence → red giant → planetary nebula → white dwarf → red dwarf → black dwarf.

Supernovae, neutron stars and black holes

Massive main-sequence stars, with masses of up to 60 times that of the Sun, do not follow the main-sequence lifecycle path – they start their death process by swelling up to form a **supergiant** star. When nucleosynthesis stops, the supergiant undergoes rapid collapse and the resulting explosion is called a **supernova**. During the supernova collapse, the energy released is so great that elements heavier than iron are formed. All elements present in the Universe, heavier than iron, were once part of a supernova explosion. What is left after the supernova explosion depends on the final mass of the supergiant. 'Low'-mass supergiant remnants form huge nebulae, containing all the gas and dust required to start stellar formation once again. 'High'-mass supergiant remnants form **neutron stars**, where the material that made up the core of the supergiant is compressed into a space with a radius of about 12 km. Many neutron stars rotate at high speed forming pulsars which emit huge 'beams' of electromagnetic radiation, such as X-rays and gamma rays, as they do so.

The 'super-high' mass supergiant remnants are called **black holes**, formed from the cores of huge stars with a core mass about 10 times the mass of the Sun. These objects are compressed into a space with a radius of about 30 km and the gravitational attraction of a black hole is so large that not even light can escape – hence the name 'black hole'. Figure 14.4 summarises the death paths of stars.

A **supergiant** is the first stage in the death of a massive main-sequence star.

A **supernova** is a gigantic explosion caused by runaway fusion reactions.

A **neutron star** is a very small dense star made out of neutrons.

A **black hole** is the most concentrated form of matter from which not even light can escape.

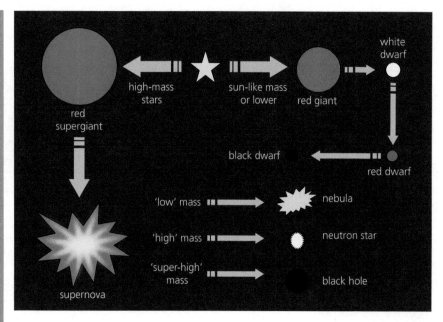

Figure 14.4 The death of stars.

Now test yourself

7 What is a supernova?
8 State the stages in the death of the Sun.
9 What are the similarities and differences between a neutron star and a black hole?

Answers on page 123

Summary

- The Solar System consists of one star (the Sun), eight planets, several dwarf planets and many moons.
- Our local patch of space is called the Solar System, which is inside a galaxy called the Milky Way, which is part of a group of galaxies called the Local Group, which, in turn, is part of a cluster of 'groups' called the Virgo Supercluster.
- We need a range of distance scales when discussing the Universe: from the scale of planets and the Solar System, where comparisons to the Earth and the Sun are best; to the Milky Way galaxy and the observable Universe, where the distance that light travels in 1 year, called a light-year, is the best unit to use. 1 ly = 9.47×10^{15} m.
- The astronomical unit, AU, is the mean distance of the Earth from the Sun. 1 AU = 1.5×10^{11} m.
- Stars form from the gravitational collapse of nebulae. Protostars form out of high-density regions inside the nebula, before forming

main-sequence stars. Main-sequence stars, with a mass similar to that of the Sun, form red-giant stars before collapsing in on themselves, forming a planetary nebula and a white dwarf. Larger mass stars form super-giants before collapsing and exploding as a supernova, leaving a nebula, neutron star or black hole.
- The stability of stars depends on a balance between gravitational force and a combination of gas and radiation pressure; stars generate their energy by the fusion of increasingly heavier elements.
- Stellar material (including the heavy elements) is returned into space during the final stages in the lifecycle of giant stars.
- The Solar System was formed from the collapse of a cloud of gas and dust, including the elements ejected by a supernova.
- **H** • The Hertzsprung–Russell (HR) diagram displays the properties of stars, and shows the evolutionary path of a star over the course of its lifetime.

Exam practice

1 The distances between objects in space are mind boggling. Even the distances between planets in our Solar System are so enormous that it takes space vehicles from Earth a very long time to get to them, many years in some cases. For example, the space craft called New Voyager that passed Pluto in 2015 was launched from Earth in January 2006 and, despite it being the fastest vehicle that has ever been sent from Earth, it took over 9 years to reach Pluto. Fortunately, there is one thing that travels so fast that we can express the vast distances of space in terms of how far it travels in 1 second or, for huge distances, the distance it travels in 1 year. This is of course light. Light travels 300 000 kilometres in 1 second and even at that speed light takes 500 s to travel to us from the Sun. We could say that the Sun is 500 light seconds away. The nearest star to our Sun is about 4 light years away, others are even millions of light years away from us.

Some of the distances used in astronomy are:

● 1 astronomical unit (AU) is the distance between the Earth and the Sun
● 1 light second is the distance travelled by light in 1 second = 300 000 km

(a) (i) What is a meant by a light-year? [1]

(ii) The Sun is 500 light seconds away from Earth.
Use the equation below to calculate this distance in km. [2]

distance = speed × time

(iii) The radius of Saturn's orbit is 9 AU. Use your answer to part (ii) to calculate the radius of its orbit in km. [2]

(b) When main-sequence stars come to the end of their 'lives' they go through stages which depend on their mass. Choose words or phrases from the list to complete the diagram that follows. [4]

red giant   black dwarf   white dwarf   supernova   neutron star

WJEC GCSE Physics Unit 2: Forces, space and radioactivity Foundation Tier SAM

2 Our Solar System is made up of eight planets, each of which may or may not have a moon or moons orbiting them, many asteroids and some dwarf planets. The planets orbit the Sun. For planet Earth the orbit time is one year. The number of years that a planet takes to orbit the Sun depends on its distance from the Sun in the way shown in the table below. The table gives data on six of the planets in the Solar System.

Planet	Mean distance from the Sun (× 10⁸ km)	Mean surface temperature (°C)	Time for one orbit of the Sun (years)
Venus	1.1.0	480	0.62
Earth	1.50	22	1.00
Mars	2.25	−23	1.88
Jupiter	7.80	−150	11.86
Saturn	14.00	−180	29.46
Uranus	29.00	−210	84.01

The graph in Figure 14.5 shows how the orbital speed of the planets changes with their distance from the Sun.

→

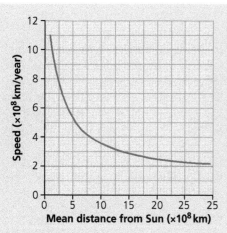

Figure 14.5

Use data from the table and graph to answer the following questions.

(a) What is the orbital speed of Saturn (in km/year)? [1]

(b) A dwarf planet, Ceres, is 700 km in diameter and has an orbital speed of 5.8×10^8 km/year. It travels 2.67×10^9 km in making one orbit of the Sun.
   (i) Use the graph to find the distance of Ceres from the Sun. [1]
   (ii) Use the equation below to calculate the orbital time of Ceres. [2]

$$\text{time} = \frac{\text{distance}}{\text{speed}}$$

(c) Estimate the mean temperature on Ceres, show your working or explain how you arrived at your answer. [2]

(d) State two reasons why Ceres takes longer than Earth to complete one orbit of the Sun. [2]

WJEC GCSE Physics Unit 2: Forces, space and radioactivity Higher Tier SAM Q4

3 The Solar System consists of the Sun and its planets.
   (a) Name the force that keeps the planets in orbit around the Sun. [1]
   (b) (i) Apart from the Earth, name one planet that has a rocky structure. [1]
       (ii) Name two planets that have a gas structure. [1]
   (c) The table below gives data on four planets in the Solar System.
       The asteroid belt lies between Mars and Jupiter. Asteroids are bits of rock, of varying size, which never collected to form a planet.
       If a planet had formed from the bits of rock, use the data in the table to estimate its:
       (i) distance from the Sun
       (ii) orbit time
       (iii) surface temperature. [3]

Planet	Mean distance from the Sun ($\times 10^8$ km)	Mean surface temperature (°C)	Time for one orbit of the Sun (years)
Earth	1.50	22	1.00
Mars	2.25	−23	1.88
Jupiter	7.80	−150	11.86
Saturn	14.00	−180	29.46

WJEC Physics P1 Higher January 2007 Q3

4 Our Sun was created and will eventually die over billions of years. The sentences below describe the stages in its life.
   A The Sun goes through a stable state.
   B The Sun shrinks to become a white dwarf.
   C Gravity pulls dust and gas together.
   D The Sun becomes a red giant.
      (a) Put the letters A, B, C and D in the correct order. [3]
      (b) In which stage, A, B, C or D, is the Sun at present? [1]

WJEC GCSE Physics P3 Foundation Tier Summer 2010 Q1

Answers and quick quiz 14 online

ONLINE

Exam practice answers and quick quizzes at www.hoddereducation.co.uk/myrevisionnotes

15 The Universe

Measuring the Universe

The first real measurements of the scale of the Universe were carried out using a technique called stellar spectroscopy by an astronomer called Edwin Hubble in 1929. Hubble knew that hot gases absorbed and emitted light with very specific wavelengths (and colours), uniquely characteristic of the elements making up the gas – like a fingerprint for each element! During the nineteenth century, astronomers had discovered that they could use stellar spectroscopy to determine the composition of stars. The spectra of all stars contain black lines, where wavelengths have been removed from the continuous spectrum by elements that make up the star – this is called an absorption spectrum. By comparing these absorption spectra to the spectra of different elements here on Earth, astronomers can tell which elements are present in the star.

Hubble was the first person to use stellar spectroscopy to measure the speed of galaxies away from Earth using a phenomenon known as redshift. The absorption spectra of many stars appeared to be displaced or 'shifted' to slightly longer wavelengths (towards the red end of the spectrum, hence 'redshift') due to the stars in the galaxies moving away from the Earth.

Hubble's law

Hubble surveyed many galaxies and plotted them against their distance away from Earth. He discovered that there was a relationship between the speed of the galaxy and its distance away from Earth, now known as Hubble's law:

> 'The speed of recession is proportional to the distance of the galaxy away from Earth.'

or

> 'The increase in redshift is proportional to the distance away from Earth.'

Hubble was also the first person to account for this pattern in **cosmological redshift**. Hubble's theory was that the increase in redshift with distance was due to the expansion of the Universe that has occurred since the Big Bang – as the Universe expands, so the wavelength of the radiation is stretched!

> **Cosmological redshift** is the change in wavelength of radiation since it was emitted due to the expansion of the Universe.

> **Exam tip**
>
> Absorption spectra questions usually involve interpreting spectral diagrams. Generally, the y-axis of these diagrams is the wavelength of the absorption lines, with the short wavelength, blue/violet lines/colours on the left and the longer wavelength, red colours/lines on the right.

1 State Hubble's law.
2 In 1842, the philosopher Auguste Comte commented that we could measure the distance and motion of planets and stars but we could never know anything about their composition. Twenty-eight years earlier, the German scientist Fraunhofer had noticed dark lines in the spectrum of the Sun. Astronomers would later use these lines to prove the philosopher incorrect. Figure 15.1 shows (in grey) the spectrum of the Sun with these 'Fraunhofer lines' and a wavelength scale.

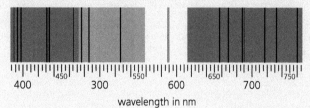

wavelength in nm

Figure 15.1

(a) Explain how the Fraunhofer lines are formed and how they tell us about the composition of the Sun.
(b) An astronomer observed the spectra of two newly discovered galaxies. It was seen that the lines in the spectra from both galaxies were 'red-shifted' when compared with the spectrum of a laboratory light source. Figure 15.2 shows the same part of the spectrum from the three sources described above. What could the scientists deduce about the distance of these two galaxies from our own? Explain your answer.

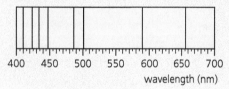

Figure 15.2

3 Figure 15.3 shows dark lines seen on the visible spectrum of a star.

400 450 500 550 600 650 700
wavelength (nm)

Figure 15.3

(a) Copy and complete the table below. Identify the elements present in the star by putting a Y (yes) or N (no) in the last column of each row.

Element	Wavelength (nm)	Present in the star?
Helium	447 502	
Iron	431 467 496 527	
Hydrogen	410 434 486 656	
Sodium	590	

(b) Explain how and why these dark lines would appear in different positions in the spectrum of a star in a distant galaxy.

Answers on page 123

The Big Bang

The Big Bang theory of the formation and subsequent evolution of the Universe was proposed as a way of explaining Hubble's measurements and his law. If the Universe began with a huge explosion, then it should still be expanding today. Cosmological redshift shows us that the rate of expansion is increasing; that is, the expansion of the Universe is 'accelerating'. The Big Bang theory also predicts that enormous amounts of energy, in the form of high-energy gamma rays, would have been produced at the moment of the Big Bang. As the Universe expanded, stretching the fabric of space, so the wavelength of the gamma rays was stretched as well – the cosmological redshift. Over 13.75 billion years of expansion, the wavelength has been stretched so much that the background remnant of these gamma rays now has the wavelength of microwaves – the Cosmic Microwave Background Radiation (CMBR), discovered by accident by Arno Penzias and Robert Wilson in 1964. They found that in whatever direction they pointed their microwave telescope, they always picked up the same background signal. They quickly realised that these signals were the remnant of the gamma rays produced at the moment of the Big Bang, cosmologically red-shifted to microwave wavelengths.

> **Exam tip**
>
> A common misconception is that the wavelength of light is increased due to the motion of the stars and galaxies that emit the light as they accelerate away from Earth. This is not the case. The increase in wavelength is caused by the expansion of space – a cosmological expansion. As the space expands, so does the wavelength.

Now test yourself

4 What is cosmological redshift evidence for?
5 The CMBR gives evidence for the Big Bang theory. How does it do this?

Answers on page 123

Summary

- Atoms of a gas absorb light at specific wavelengths, which are characteristic of the elements in the gas.
- You can use data about the spectra of different elements to identify gases from an absorption spectrum.
- Scientists in the nineteenth century were able to reveal the chemical composition of stars by studying the absorption lines in their spectra.
- Edwin Hubble's measurements on the spectra of distant galaxies revealed that the wavelengths of the absorption lines are increased and that this 'cosmological redshift' increases with increasing distance.

- The cosmological redshift of the radiation emitted by stars and galaxies is due to the expansion of the Universe since the radiation was emitted.
- The Big Bang theory of the origin of the Universe predicted the existence of background radiation, which was subsequently detected accidentally in the 1960s, and that the Cosmic Microwave Background Radiation (CMBR) is the red-shifted remnant of radiation from the origin of the Universe.
- Cosmological redshift and the CMBR have provided the evidence for the establishment of the Big Bang theory of the origin of the Universe.

Exam practice

1 Figure 15.4 shows gas atoms in the cooler, outer atmosphere of a star in a galaxy that is 20 million light years away.
  (a) Write down the time taken for light to get to us from the star. [1]
  (b) Figure 15.4 shows light from the inner part of the star passing through its outer atmosphere. Some of the wavelengths are absorbed.
  Figure 15.5 shows three spectra.

→

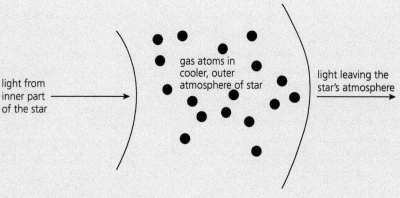

Figure 15.4

diagram A

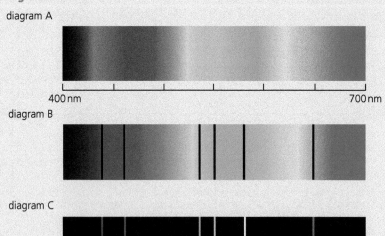

400 nm   700 nm

diagram B

diagram C

Figure 15.5

Put a tick (✔) in the correct column in the table to show which is the correct diagram for each statement. [2]

	Diagram A	Diagram B	Diagram C
Light leaving the star's atmosphere			
Light from inner part of the star			

(c) The spectrum in Figure 15.6 is from another star in the same galaxy. Explain how the lines show that the stars are different. [2]

Figure 15.6

(d) Lines in spectra from distant galaxies are shifted towards the red end of the spectrum.
  (i) State what has happened to the wavelengths of those lines. [1]
  (ii) The lines in spectra from some other galaxies are further red-shifted. State what this tells us about those galaxies. [1]

(e) Cosmological redshift supports the Big Bang theory. Name one other piece of evidence that supports this theory. [1]

WJEC GCSE Science A/Physics P1 Foundation Tier January 2016 Q3

2 (a) Explain how scientists in the nineteenth century were able to reveal the chemical composition of stars. [3]

→

(b) Describe the findings arising from Sir Edwin Hubble's measurements on the spectra of distant galaxies. [2]

(c) Explain how the presence of Cosmic Microwave Background Radiation (CMBR) supports the Big Bang theory. [2]

WJEC GCSE Science A/Physics P1 Foundation Tier January 2016 Q5

3 An absorption spectrum from a star is a pattern of black lines on a coloured background.

Figure 15.7

(a) The boxes in the left column below list four features of the spectrum. The boxes in the right column list the causes of these features. Draw a line from each feature on the left to its correct cause on the right. [3]

Feature	Cause
A single black line	Due to wavelengths of visible light emitted by the star
The black lines move towards the red end of the spectrum	Due to the gas elements in the star
Pattern of black lines	Due to cosmological redshift
Coloured background	Due to one wavelength of light being absorbed

(b) Name the theory that is supported by cosmological redshift. [1]

WJEC GCSE Science A/Physics P1 Foundation Tier June 2016 Q1

4 The first diagram in Figure 15.8 shows the spectrum of white light after it is passed through hydrogen in the laboratory.

The second spectrum comes from a galaxy that is 4×10^9 light years away.

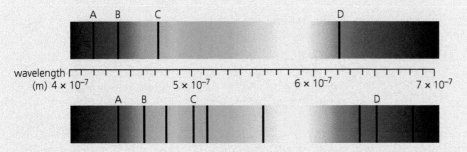

Figure 15.8

(a) State the time taken for light to travel from the galaxy to us. [1]

(b) Calculate the change in wavelength of the line A between the laboratory spectrum and the galaxy's spectrum. [1]

(c) Explain what information can be obtained about the galaxy by comparing the two spectra. (Do not include in your answer the development of theories of the Universe.) [6 QER]

WJEC GCSE Science A/Physics P1 Higher Tier June 2016 Q6

Answers and quick quiz 15 online

ONLINE

16 Types of radiation

Inside the nucleus

REVISED

The nucleus of an atom contains positively charged particles, protons, and neutral particles, neutrons. The number of protons in the nucleus is called the **proton number**, Z; the number of protons plus the number of neutrons is called the **nucleon number**, A. The values of Z and A are often shown using the $_Z^A X$ notation, where X is the chemical symbol for the atom in question. For example, 52.4 per cent of all naturally occurring lead atoms have nuclei made up of 82 protons and 126 neutrons, a total of 208 nucleons, i.e. $_{82}^{208} Pb$. Lead also has other **isotopes** – nuclei with the same number of protons, but different numbers of neutrons. The different isotopes are often written as Pb-208, Pb-207, etc., where the number refers to the nucleon number.

> **Proton number** is the number of protons.
>
> **Nucleon number** is the number of protons and neutrons.
>
> **Isotopes** are different forms of a particular element. Isotopes have the same number of protons but different numbers of neutrons.

Now test yourself

TESTED

1 Use the $_Z^A X$ notation to describe the following radioactive nuclei in the table.

Nucleus	Proton number, Z	Number of neutrons	$_Z^A X$
Lithium	3	4	
Carbon	6	7	
Strontium	38	52	
Technetium	43	56	

Answers on page 123

> **Exam tip**
>
> The proton number is sometimes called the atomic number, and the nucleon number is sometimes called the atomic mass. These values are not exact, because they describe atoms not nuclei.

> **Exam tip**
>
> You can easily work out the number of neutrons in a nucleus by subtracting the proton number from the nucleon number.

Nuclear radiation

REVISED

Some types of atom are radioactive. This means that the nucleus of the atom is unstable, due to an imbalance of protons and neutrons, and it can break apart, emitting ionising radiation in the form of alpha (α), beta (β) or gamma (γ) radiation.

- Alpha particles are helium nuclei. They are the most ionising (and, therefore, cause the most harm inside the body to living cells) and the least penetrating type of nuclear radiation – they are absorbed by a thin sheet of paper or by skin. Alpha-emitting nuclear waste is easily stored in plastic or metal canisters.
- Beta particles are high-energy electrons. They have medium ionising ability (and cause little harm inside the body to living cells); they

are absorbed by a few millimetres of aluminium or Perspex plastic, and beta-emitting nuclear waste is stored inside metal canisters and concrete silos.

● Gamma rays are high-energy electromagnetic waves. They are the least ionising (about 20 times lower than alpha particles) and, therefore, cause the least harm to living cells inside the body. They are the most penetrating, able to travel through several centimetres of lead. Nuclear waste containing gamma emitters needs storage inside lead-lined, thick concrete silos.

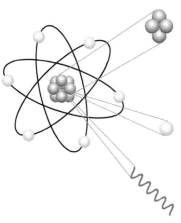

Alpha (α) radiation
These are particles, not rays, and travel at about 10% of the speed of light. An α particle is identical to a helium nucleus, consisting of 2 protons and 2 neutrons joined together.

Beta (β) radiation
These are fast-moving electrons that come from the nucleus. They travel at about 50% of the speed of light.

Gamma (γ) radiation
This is an electromagnetic wave. It travels at the speed of light (3×10^8 m/s). It has very high energy.

Figure 16.1 Alpha, beta and gamma radiation.

Now test yourself — TESTED

2 Complete the summary table of the three different types of ionising radiation.

Type of radiation	Symbol	$^A_Z X$	Penetrating power	Ionising power
Alpha				
Beta				
Gamma				

Answers on page 124

Stability of the nucleus — REVISED

In a stable atom there is an optimum balance between the number of protons and the number of neutrons in the nucleus. But some isotopes can have too few or too many neutrons, causing an imbalance and making the nucleus unstable and radioactive. For example:

● the lead isotope Pb-181 has only 99 neutrons and is an alpha particle emitter
● the lead isotope Pb-214 has 132 neutrons and is a beta emitter. The nucleus becomes more stable by emitting alpha or beta particles, restoring the optimum balance of protons and neutrons, and sometimes emitting gamma radiation too.

Nuclear equations

REVISED

The ^A_ZX notation can be used to represent the decay of radioactive nuclei in a nuclear equation. Alpha particles are written as $^4_2$He because they consist of two protons and two neutrons, like a helium nucleus. Beta particles are written as $^0_{-1}$e, because they are electrons.

Alpha decay

The nuclear decay equation for the alpha decay of lead-181 is:

$$^{181}_{82}\text{Pb} \rightarrow ^4_2\text{He} + ^{177}_{80}\text{Hg}$$

The lead-181 nucleus emits an alpha particle, losing 4 nucleons (2 protons + 2 neutrons), forming mercury-177.

Beta decay

The nuclear decay equation for the beta decay of lead-214 is:

$$^{214}_{82}\text{Pb} \rightarrow ^0_{-1}\text{e} + ^{214}_{83}\text{Bi}$$

The lead-214 nucleus emits a beta particle (electron). The nucleon number stays the same, but the proton number goes up by one, forming bismuth-214.

> **Exam tip**
>
> All nuclear equations must balance: the total proton number on each side must be the same, and the total nucleon number on each side must be the same.

Now test yourself

TESTED

3 Complete the following nuclear decay equations, by determining A and Z:

(a) $^{241}_{95}\text{Am} \rightarrow ^4_2\text{He} + ^A_Z\text{Np}$

(b) $^{225}_{88}\text{Ra} \rightarrow ^0_{-1}\text{e} + ^A_Z\text{Ac}$

Answers on page 124

Background radiation

REVISED

Background radiation is all around us and it comes naturally from our environment and from artificial (human-made) sources. The background-radiation count rate needs to be subtracted from any measurements made of nuclear radiation. The pie chart in Figure 16.2 shows the mean contribution of the different sources to our background radiation.

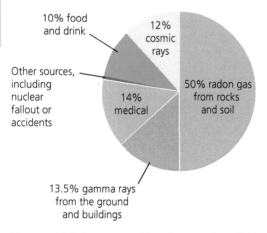

Figure 16.2 Sources of background radiation.

Most of the background radiation comes from naturally occurring sources, primarily from the ground, rocks and from space. Artificial background radiation mostly comes from medical sources, predominantly as a result of medical and dental examinations using X-rays. The largest source of the background radiation comes from the radioactive element radon, emitted from rocks (granite) and the soil. Radon is a gas, and it can escape from granite and is easily breathed in by humans, entering our lungs, where it can decay. The alpha particles emitted by decaying radon are absorbed by the cells lining the lungs, causing the cells to die or mutate (forming cancers).

Monitoring and measuring radioactive decay

REVISED

Radioactive decay is a random process and this has consequences when undertaking experimental work. Readings should be repeated and mean averages taken; background radiation must be subtracted from the readings, and the readings generally need to be taken over a lengthy period. All of these techniques are used to reduce the effect of random fluctuations in the measurements.

Now test yourself

TESTED

4 What is background radiation?
5 How are radioactivity measurements improved by taking into account the fact that they are random processes?

Answers on page 124

Nuclear waste

REVISED

High-level nuclear waste from nuclear reactor cores is stored temporarily in water, surrounded by lots of concrete and lead shielding, while it cools. The radiation is absorbed by the water, concrete and lead. In the longer term, this waste is encased in blocks of glass – a process known as 'vitrification'. The glass blocks are then stored deep underground, where the surrounding rocks absorb the radiation. The initial cooling and vitrification processes take decades and the radioactive material stays radioactive for millions of years. Intermediate nuclear waste from medical processes is less radioactive than high-level waste and is mixed with concrete and poured into steel drums, before being stored securely underground.

> **Exam tip**
>
> Question 6 is an example of a question containing complex tables. In the exam you must spend time studying tables very carefully. Re-read the column and row headings and make sure that you know exactly what data is being given to you.

Now test yourself

TESTED

6 Some radioactive elements emit more than one type of radiation.
   The apparatus in Figure 16.3 was used to investigate the radiation emitted from three sources, A, B and C. The sources were always placed at the same position, close to the detector. The table below shows the mean counts per minute obtained when different materials were placed between the sources and the detector. All the readings have been corrected for background radiation.

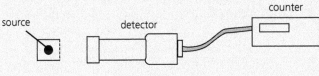

Figure 16.3

	Mean counts/min with nothing between source and detector	Mean counts/min with thin paper in the way	Mean counts/min with 3 mm of aluminium in the way	Mean counts/min with 2 cm of lead in the way
A	256	256	256	85
B	135	80	80	0
C	310	310	188	0

(a) How can you tell that source A is emitting gamma (γ) radiation?
(b) Which source, A, B or C, emits alpha (α) particles? Give a reason for your answer.
(c) The beta radiation source contains atoms of strontium-90, $^{90}_{38}$Sr.
  (i) Explain what happens to a Sr-90 atom when it decays.
  (ii) The elements nearest to strontium in the periodic table are shown below. Use the information in the table to determine the daughter nucleus produced when Sr-90 decays by beta emission.

Element	Krypton	Rubidium	Strontium	Yttrium	Zirconium
Symbol	Kr	Rb	Sr	Y	Zr
Proton number	36	37	38	39	40

Answers on page 124

Summary

- Substances that are radioactive can emit alpha (α; $^4_2$He), beta (β; $^0_{-1}$e) and gamma (γ) radiation.
- Alpha (α), beta (β) and gamma (γ) radiation, ultraviolet light and X-rays are all types of ionising radiation. Ionising radiation is able to interact with atoms and damage cells because of the energy it carries.
- Radioactive emissions from unstable atomic nuclei arise because of an imbalance between the numbers of protons and neutrons.
- The number of protons and neutrons in an atomic nucleus is called the nucleon number or the mass number, (A), and the number of protons is called the proton number, (Z); chemists usually call it the atomic number.
- The nuclear symbols in the form of the ^A_ZX notation (where X is the atomic symbol from the periodic table) are used to describe radioactive atoms, decays and balanced radioactive decay equations.
- The waste materials from nuclear power stations and nuclear medicine are radioactive; some of them will remain radioactive for thousands of years.
- When measurements of radiation are taken, an allowance for background radiation must be made.
- Alpha, beta and gamma radiation have different penetrating powers. Alpha radiation is absorbed by a thin sheet of paper, beta radiation is absorbed by a few millimetres of aluminium or Perspex, but gamma radiation is only absorbed by a few centimetres of lead.
- The differences in the penetrating power of alpha, beta and gamma radiation determine their potential for harm. Alpha radiation is easily absorbed but is the most ionising. Gamma rays are very penetrating but are about 20 times less ionising than alpha radiation.
- Radioactive waste is stored in a series of containment systems. Steel canisters, water, concrete, glass and lead are all used to shield the environment from harmful doses of radiation. The radiation produced by the waste is absorbed by the different types and thicknesses of containment.
- The long-term solution to the storage of radioactive waste is deep underground, where the harmful radiation can be absorbed by the surrounding rocks.
- Background radiation is all around us and comes from natural or artificial (human-made) sources.
- Natural sources of background radiation include radon from rocks, gamma rays from the ground and buildings, cosmic rays from space and radiation contained in food and drink.
- Artificial sources of background radiation include X-rays from medical examinations and nuclear fallout from weapons tests or accidents.
- Most of our background radiation (between 50 per cent and 90 per cent) comes from radon gas (depending on where you live). Places like Cornwall, where there is a lot of granite rock, have higher levels of radon gas because the granite contains uranium that decays (eventually) to radon.

Exam practice

1 Read the information below.
The first 92 elements in the periodic table occur naturally on the Earth. Other elements have been created by mankind, usually inside nuclear reactors. The atoms of some elements exist in different forms, which are called isotopes. Isotopes of the same element all have the same number of protons. However, isotopes of different elements may have the same nucleon number, some of which are shown in the table below.

Isotope	Proton number	Nucleon number
Americium Am)	95	238
Uranium (U)	92	238
Thorium (Th)	90	238
Californium (Cf)	98	238

(a) Read the statements below and tick (✓) the correct ones. [3]

Statement	
Atoms of all of these isotopes have the same number of protons in their nuclei.	
An atom of uranium has 92 neutrons in its nucleus.	
An atom of californium has the greatest number of protons in its nucleus.	
An atom of californium has the smallest number of neutrons in its nucleus.	
Uranium is not a naturally occurring element.	
An atom of uranium has 92 protons in its nucleus.	

(b) Complete the decay equation of uranium-238 into thorium in the equation below. [2]

$$^{238}_{92}U \rightarrow ^4_2\alpha + \, ^{\,}_{\,}Th$$

(c) Identify the two correct isotopes of uranium from the list below. [2]

$$^{238}_{92}U \qquad ^{238}_{89}U \qquad ^{234}_{90}U \qquad ^{235}_{92}U \qquad ^{238}_{91}U$$

WJEC GCSE Physics Unit 2: Forces, space and radioactivity Foundation Tier SAM Q1

2 The sources and percentages of background radiation are shown in the pie chart in Figure 16.4. The pie chart is not drawn to scale.

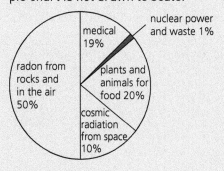

Figure 16.4

(a) (i) What percentage of background radiation comes from natural sources? [1]
   (ii) Use the pie chart to explain why an increase in nuclear-power generation would only produce a small increase in the background radiation. [1]
(b) Give a reason why background radiation varies from place to place. [1]

WJEC GCSE Physics P2 Foundation Tier Summer 2010 Q4

3 A piece of rock gives a reading on a counter which is connected to a radiation detector. The rock is wrapped in various materials and different readings in counts per minute (cpm) are observed.

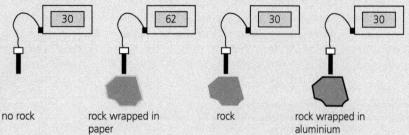

no rock   rock wrapped in   rock   rock wrapped in
                 paper   aluminium

Figure 16.5

(a) (i) State the type, or types of radiation that the rock emits.   [2]
    (ii) Explain your choice.   [2]
(b) The rock is now wrapped in lead. Explain what you would expect the reading on the counter to be.   [2]

WJEC/EDUQAS GCSE Physics Component 1 Concepts in Physics Foundation Tier Sample Paper Q8

4 (a) (i) Give a reason why radioactive waste is a health risk to the population, e.g. a cancer risk.
      (ii) Give a reason why radioactive waste is expensive to dispose of safely.   [2]
  (b) A sample of radioactive waste emits alpha (α), beta (β) and gamma (γ) radiation. When placed in front of a detector the sample gives a reading of 450 counts/minute. The diagrams in Figure 16.6 show the count rate when different absorbers are placed between the sample and the detector.

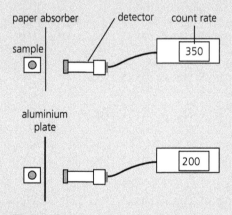

Figure 16.6

(i)   How much of the 450 counts/minute is due to alpha (α) radiation?   [1]
(ii) How much of the 450 counts/minute is due to gamma (γ) radiation?   [1]
(iii) Explain why the aluminium absorber reduces the original count rate by 250 counts/minute.   [2]

WJEC GCSE Physics P2 Foundation Tier Summer 2010 Q8

Answers and quick quiz 16 online

ONLINE

17 Half-life

Radioactive decay and probability

Radioactive decay is a random process, but all radioactive substances decay in a similar way, following the same pattern. Radioactive atoms have a constant probability of decay, and follow the same rules of probability that other random processes follow, such as throwing dice. The probability of throwing a six on a dice is 1/6, in the same way that a radioactive atom has a constant numerical probability of decay. This means that, although you cannot say for definite which atoms of a radioactive substance will decay in a set time, you can predict how many of them will decay. Each radioactive substance has a given probability of decay; some have very high probabilities and decay very quickly indeed, whereas others have extremely small probabilities and remain radioactive for very long times. The unit of radioactive decay activity is the becquerel, Bq. An activity of 1 Bq is equivalent to 1 radioactive decay per second.

Half-life

Radioactive materials decay in a very predictable way. The graph in Figure 17.1 shows a typical radioactive decay curve, for the decay of iridium-192, Ir-192.

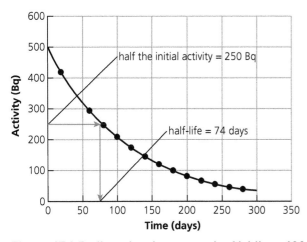

Figure 17.1 **Radioactive decay graph of iridium-192.**

The initial activity of the sample is 500 Bq. After 74 days the activity has fallen to 250 Bq – half the original amount. This time is called the half-life of iridium-192. After another 74 days, the activity has fallen to 125 Bq – half of 250 Bq – and after each half-life the activity halves. The half-life of a radioactive substance is the time taken for the activity of the substance to halve. Some substances have very short half-lives and decay extremely quickly, whereas others have very long half-lives and remain radioactive for very long times – some have half-lives considerably longer than the age of the Universe.

An iridium-192 source has a half-life of 74 days and an initial activity of 1200 Bq. What will the activity of the iridium source be after 222 days?

Answer

Number of half-lives in 222 days = $\dfrac{222 \text{ days}}{74 \text{ days}}$ = 3 half-lives

After 1 half-life (74 days) the activity will be $\dfrac{1200 \text{ Bq}}{2}$ = 600 Bq

After the second half-life (148 days), the activity will be $\dfrac{600 \text{ Bq}}{2}$ = 300 Bq

After the third half-life (222 days), the activity will be $\dfrac{300 \text{ Bq}}{2}$ = 150 Bq

Exam tip

The decay of every radioactive material follows the same pattern. The activities and the half-lives may be very different, but the shape of the decay curve is always the same.

Now test yourself

TESTED

1 Why do all radioactive substances decay with a similar pattern?
2 What is meant by radioactive half-life?
3 To study blood flow, a doctor injects some technetium-99 (Tc-99) into a patient. The gamma radiation given out by the Tc-99 atoms is detected using a gamma camera outside the patient's body. The graph in Figure 17.2 shows how the count rate from a sample of Tc-99 changes with time.

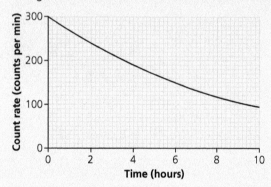

Figure 17.2

(a) (i) How many hours does it take for the count rate to fall from 300 counts per minute to 150 counts per minute?
    (ii) What is the half-life of Tc-99?
    (iii) How long will it take for the count rate to fall from 300 to 75 counts per minute?
(b) Explain why an alpha-emitting source would be unsuitable to study blood flow.

Answers on page 124

Using radioactivity

REVISED

The uses of radioactive materials depend on their properties, in particular:
● half-life
● penetrating power
● ionising power.

In medicine, radioactive materials are used in two main ways: in imaging and in therapy (treatment). The isotope carbon–14 can also be used to date the age of very old (dead) materials.

Radio-imaging

The radioactive material is injected into the body. It makes its way to the particular place that requires investigation and it emits gamma rays, which can be detected outside of the body. Gamma emitters are used, such as technetium-99 with a half-life of six hours. This will only remain radioactive for 30 hours or so (about 5 half-lives). Gamma rays cause few problems to the body as they pass straight through and are very weak ionisers.

Radiotherapy

This involves the use of radioactive materials to kill affected (usually cancerous) cells. Beta radiation may be used, as this has a short range in flesh, so that it only damages the cells close to the target area. The half-life chosen is typically a few days, but depends on the dose required – longer half-lives will deliver higher doses.

Carbon dating

Carbon-14 is a naturally occurring radioactive isotope of carbon. Only about 1 atom in every 10 000 000 000 carbon atoms is an atom of carbon-14. Carbon-14 is a radioactive beta emitter with a half-life of 5730 years and it can be used to date organic objects up to about 60 000 years old – about the time when our early ancestors, the Homo sapiens, started to migrate out of Africa. All living things contain carbon, and the ratio of non-radioactive carbon-12 atoms to radioactive carbon-14 atoms in living material is known very precisely – as it depends on the composition of the carbon dioxide in the atmosphere. When an organic living creature or plant dies, the ratio of carbon-12 to carbon-14 starts to change as the carbon-14 decays and no more fresh carbon-14 is added (because photosynthesis and/or respiration is no longer carried out by the creature or plant once it is dead). If the ratio of carbon-12 to carbon-14 in a dead organic object is measured, then it is possible to use the half-life of carbon-14 to work backwards to find out when the ratio was the same as it is in living organisms now. A graph similar to the one in Figure 17.3 can be used to measure the percentage of carbon-14 remaining, compared with a living sample.

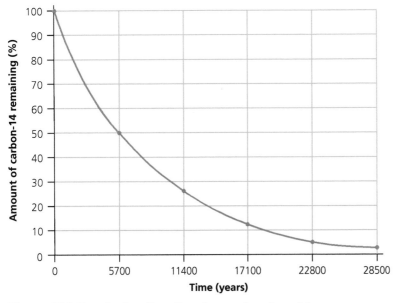

Figure 17.3 Graph of radioactive decay of carbon-14.

> **Exam tip**
>
> Question 6 below involves drawing a decay graph. When you have to draw a graph in an exam, you are nearly always given the axes and grid. The most important thing is to plot the points accurately and correctly, and draw a best-fit curve – always check your graph two or three times after you have plotted it and before you draw the curve, and use pencil so that you can rub it out if you make a mistake.

Now test yourself

4 The use of a radioactive substance depends on which three properties of the radioactive substance?
5 List two medical uses of radioactive substances.
6 (a) Carbon-14 has a half-life of 5700 years. Sketch a graph of activity against time, showing the decay of carbon-14 from an initial activity of 64 counts per minute.
   (b) While trees are alive they absorb and emit carbon-14 (in the form of carbon dioxide) so that the amount of carbon-14 in them remains constant.
      (i) What happens to the amount of carbon-14 in a tree after it dies?
      (ii) A sample of wood from an ancient dwelling gives 36 counts per minute. A similar sample of living wood has 64 counts per minute. From your graph, deduce the age of the dwelling. (Show on your graph how you obtained your answer.)

Answers on page 124

Summary

- Radioactive decay is a random event, governed by the laws of probability. Radioactive decay can be modelled by rolling a large collection of dice, tossing a large number of coins or using a suitably programmed spreadsheet.
- The half-life is the length of time it takes for half the atoms in the sample to decay; this is a constant for a particular element. Half-lives vary between different radioactive elements from less than a second to billions of years.
- The activity of a sample of isotope can be plotted against time on a graph, and from this the half-life of that isotope can be measured. The graph is called a radioactive decay graph.

- The unit of radioactive decay is the becquerel, Bq. 1 Bq is 1 radioactive decay per second.
- Carbon-14 is a naturally occurring radioactive isotope of carbon. It emits beta particles with a half-life of 5370 years. By measuring the proportion of carbon-14 to the usual isotope carbon-12, organic objects up to around 60000 years old can be dated.
- The characteristics of the different forms of radioactive decay, such as half-life, penetrating power and ionising ability, mean that they are useful for different purposes, for example medical uses such as radio-imaging and radiotherapy.

Exam practice

1 A student does an experiment with dice to investigate radioactive decay. The dice, which represent radioactive atoms, are thrown together onto the floor. Those that show a six are removed. These represent the atoms whose nuclei have decayed. The remaining dice (undecayed atoms) are thrown again and the process is repeated several times. The student starts with 600 dice.
   (a) (i) Predict how many of the dice would show a six on the first throw. [1]
      (ii) State why the student cannot predict which dice will show a six. [1]
   (b) The results of the experiment are shown in the table below.

Throw	Number of sixes	Number of dice remaining
0	0	600
1	95	505
2	85	420
3		350
4	60	290
5	50	240
6	40	200
7	30	170
8	25	145

   (i) Fill in the gap in the table above. [1]
   (ii) Plot the results on the grid below and draw a suitable line. Three points have been plotted for you. [3]

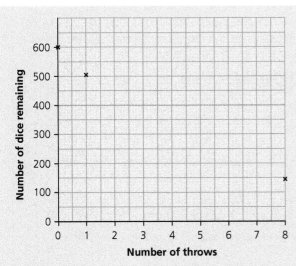

Figure 17.4

(iii) Draw lines on your graph to enable you to find the half-life of the dice. [2]

(c) Americium-241 is a radioactive substance which is used in smoke alarms in houses. It decays by emitting alpha particles.
    (i)   State why Americium-241 is radioactive. [1]
    (ii) What is an alpha particle? [1]
    (iii) Explain why the use of Americium-241 in house smoke alarms, when in normal use, does not present a significant health risk to people living in the houses. [2]

WJEC GCSE Additional Science/Physics P2 Higher Tier January 2016 Q1

2 The graph in Figure 17.5 shows the radioactive decay in counts per minute (cpm) of a sample of carbon-14.

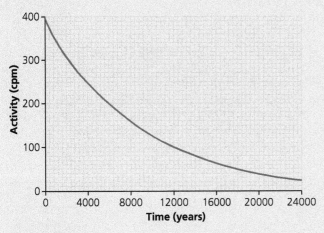

Figure 17.5

(a) Use information from the graph to answer the following questions.
    (i)   State the activity after 4000 years. [1]
    (ii) State the time taken for the activity to fall from 400 cpm to 100 cpm. [1]
    (iii) State the half-life of carbon-14. [1]
    (iv) State the time it would have taken for the activity to have fallen from 800 cpm to 400 cpm. [1]
(b) The nuclear symbol for carbon-14 is C. Complete the following table for the nucleus of carbon-14. [3]

Nucleon number	
Number of protons in its nucleus	
Number of neutrons in its nucleus	

WJEC GCSE Additional Science/Physics P2 Higher Tier May 2016 Q4

3 Nuclear medicine uses radioisotopes which emit radiation from within the body. One tracer uses iodine, which is injected into the body to treat the thyroid gland. The table shows four isotopes of iodine.

Form of iodine	Radiation emitted	Half-life
Iodine-125	Gamma	59.4 days
Iodine-128	Beta	25 minutes
Iodine-129	Beta and gamma	15 000 000 years
Iodine-131	Beta and gamma	8.4 days

(a) Iodine-129 emits both beta and gamma radiation. Describe the nature of these types of radiation. [2]

(b) The table shows that the half-life of iodine-125 is 59.4 days. State what this means. [2]

(c) (i) Use the data to explain why iodine-131 is the most suitable form of iodine for treating thyroid cancer. [2]

(ii) Patients are advised that after treatment with iodine-131, the radiation they are exposed to will not drop to the background value until 12 weeks after treatment. Calculate the fraction of radioactivity due to iodine-131 remaining after 12 weeks. [3]

WJEC GCSE Additional Science/PhysicsP2 Higher Tier May 2016 Q3

4 Isotopes of iodine can be used to study the thyroid gland in the body. A small amount of the radioactive isotope is injected into a patient and the radiation is detected outside the body. Three isotopes that could be used are: $^{123}_{53}I$; $^{131}_{53}I$ and $^{132}_{53}I$. They have half-lives of 13.22 hours, 8 days and 13.2 hours respectively.

Answer the following question in terms of the numbers of particles.

(a) Compare the structures of the nuclei of $^{123}_{53}I$ and $^{131}_{53}I$. [2]

(b) The nucleus of $^{131}_{53}I$ decays into xenon (Xe) by giving out beta (β) and gamma (γ) radiation.

(i) What is beta radiation? [1]

(ii) Complete the equation below to show the decay of I-131. [2]

$$^{131}_{53}I \rightarrow \ ^{...}_{54}Xe + \ ^{0}_{...}\beta + \gamma$$

(c) The isotope $^{123}_{53}I$ decays by gamma emission. Explain why it is better to use $^{123}_{53}I$ than $^{131}_{53}I$ as a medical tracer. [2]

(d) (i) I-131 has a half-life of 8 days. Explain what this statement means. [2]

(ii) Following the nuclear power-station disaster in Japan 2011, people living in the area were given non-radioactive iodine-127 ($^{127}_{53}I$) supplement tablets to reduce their intake of iodine-131 that leaked from the reactor. Calculate the length of time that people had to take the supplement before the activity of iodine-131 reduced to approximately 3 per cent of its original value immediately after the leak. [2]

WJEC GCSE Physics Unit 2: Forces, space and radioactivity Higher Tier SAM Q7

Answers and quick quiz 17 online

ONLINE

18 Nuclear decay and nuclear energy

Nuclear fission

The stability of an atomic nucleus is dictated by the number of protons and neutrons within the nucleus. Heavy nuclei (generally with atomic numbers above 27 such as iron, Fe) tend to have a large number of neutrons compared to the number of protons. This makes them unstable and they can break apart – a process called nuclear fission. Nuclei that can undergo fission are called fissile nuclei. Nuclear fission releases energy. On a large scale, huge amounts of energy can be produced in a controlled way in a nuclear reactor. The energy (in the form of heat) can be used to generate electricity – known as nuclear power. Fission of 1 kg of nuclear fuel can produce 83 000 000 000 000 J of energy; by comparison, combustion of 1 kg of coal can produce 35 000 000 J. In one type of nuclear reactor, uranium-235 nuclei are broken up into two daughter nuclei when bombarded by slow-moving neutrons. The process produces two or three more neutrons, which in turn can induce the fission of other U-235 nuclei, and so on, starting a sustainable chain reaction.

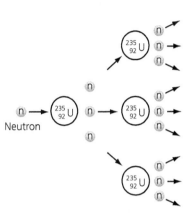

Figure 18.1 A chain reaction in uranium-235.

We can represent one fission event by the nuclear equation:

$$^{235}_{92}U + ^{1}_{0}n \rightarrow ^{144}_{56}Ba + ^{89}_{36}Kr + 3^{1}_{0}n + \text{energy}$$

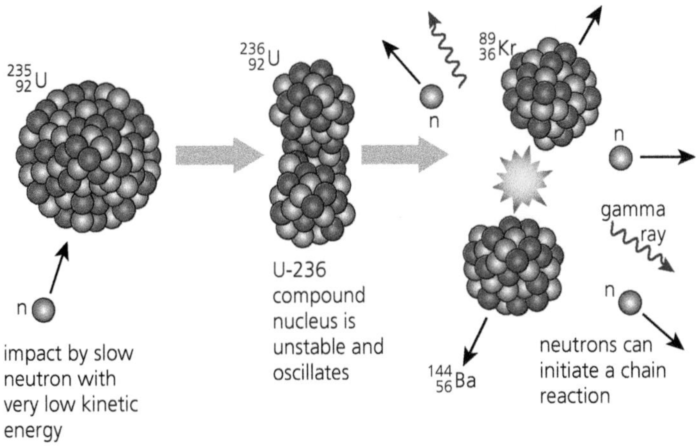

fission yields fragments of intermediate mass, an average of 2.4 neutrons, and high kinetic energy

$^{235}_{92}U$

$^{236}_{92}U$

$^{89}_{36}Kr$

n

gamma ray

$^{144}_{56}Ba$

n

impact by slow neutron with very low kinetic energy

U-236 compound nucleus is unstable and oscillates

neutrons can initiate a chain reaction

Figure 18.2 Fission of uranium-235.

The daughter nuclei of fission reactions are themselves radioactive and decay by alpha, beta or gamma emission. They have wide-ranging but generally extremely long half-lives (typically hundreds of thousands of years) and they will remain dangerously radioactive for a very long time. This why some nuclear waste requires extremely secure long-term storage.

Exam tip

Questions involving nuclear fission can be tricky. The number of neutrons produced when U-235 undergoes fission is not constant and varies between one and three. Always read the question carefully and study any fission diagrams or equations to make sure you are certain how many neutrons are produced.

Controlling and containing the reaction

Fission only occurs if the bombarding neutrons are moving slowly enough. To initiate a chain reaction, the fast-moving neutrons released in the fission events need to be slowed down. The fuel rods in the reactor are surrounded by a material called a moderator, which slows the neutrons down. This is normally water (which also acts as the coolant and the mechanism of heat transfer for the reactor). The chain reaction can be completely stopped, speeded up or slowed down by controlling the number of slow neutrons in the reactor. This is achieved by inserting neutron-absorbing control rods into spaces between the fuel rods. A nuclear reactor core emits huge amounts of radioactivity in the form of gamma rays and neutrons, and needs to be shielded from the environment. There is a thick metal casing around the reactor vessel and also a thick concrete containment structure surrounding that.

Now test yourself

TESTED

1 What is nuclear fission?
2 How many times more energy is produced by the fission of 1 kg of uranium-235 compared to the combustion of 1 kg of coal?
3 In an uncontrolled nuclear fission reaction, when a slow-moving neutron strikes an atom of U (uranium), the atom splits. In this reaction two fast-moving neutrons are produced together with the radioactive fission fragments of Ba (barium) and Kr (krypton).
   (a) Write out and complete the nuclear equation for this reaction.

$$^{235}_{92}U + ^{1}_{0}n \rightarrow ^{144}_{...}Ba + ^{...}_{36}Kr + 2^{1}_{0}n + energy$$

   (b) In a nuclear reactor, the fission reaction is controlled using control rods of boron steel, which readily absorbs neutrons, and a graphite moderator which improves the chances of uranium atoms splitting apart.
      (i) State how the graphite moderator improves the possibility of fission of uranium.
      (ii) Explain how the energy released from a nuclear reactor can be increased.

Answers on page 124

Nuclear fusion

REVISED

The energy produced by our Sun is a result of nuclear reactions. In this case, the nuclear reaction involves the fusion (joining together) of light nuclei, such as the isotopes of hydrogen: deuterium, $^{2}_{1}H$, and tritium, $^{3}_{1}H$, and the release of energy.

In a prototype nuclear fusion reactor here on Earth, an ionised gas (plasma) of two isotopes of hydrogen: deuterium, $^{2}_{1}H$, and tritium, $^{3}_{1}H$, are fused together at very high temperatures (many millions of degrees Celsius).

$$^{2}_{1}H + ^{3}_{1}H \rightarrow ^{4}_{2}He + ^{1}_{0}n$$

Fusion of 1 kg of hydrogen could produce over 7 times as much energy as fission of 1 kg of uranium-235. The problem with designing nuclear fusion reactors is containing the high temperature plasma and also containing the radiation emitted during the process.

> **Exam tip**
>
> Nuclear fission and fusion questions frequently involve completion of nuclear equations. These must balance: the total proton number on each side must be the same, and the total nucleon number on each side must be the same.

Now test yourself

4 What is nuclear fusion?
5 Why will any nuclear fusion reactors built here on Earth require large amounts of concrete shielding?
6 The following equation shows a nuclear reaction. This reaction only takes place if the particles on the left-hand side of the equation move very quickly towards each other. This needs a very high temperature. The reaction then releases a huge amount of energy.

$$_1^2H + {}_1^2H \rightarrow {}_2^3He + {}_0^1n$$

(a) Choose the correct word or words in the brackets of each sentence below.
   (i) The particles that collide together in this reaction are atoms of [hydrogen / helium / oxygen].
   (ii) This is an example of a [fission / chain / fusion] reaction.
(b) Give two reasons why this reaction is very difficult to control.
(c) Outline the advantages of producing electricity from nuclear fusion rather than nuclear fission in the future.

Answers on page 124

Summary

- The absorption of slow neutrons can induce fission of uranium-235 nuclei, referred to as fissile nuclei, releasing energy, and the emission of neutrons from such fission can lead to a sustainable chain reaction.
- Moderator material in a nuclear reactor acts to slow down the fast-moving neutrons produced by the nuclear fission process, so that they can cause further fission.
- Control rods are neutron-absorbing rods that can be moved up and down to control the number of slow neutrons there are inside the fuel rods.
- Most of the decay products of nuclear fission are radioactive, many of them with very long half-lives, so they have to be carefully stored within the containment structure of the nuclear reactor.
- High-energy collisions between light nuclei, especially isotopes of hydrogen, can result in fusion which releases enormous amounts of energy.
- For fusion to take place, very high temperatures are required, which are difficult to achieve and control.
- The problems of containment in fission and fusion reactors also include neutron and gamma shielding, and pressure containment in fusion reactors.

Exam practice

1 One possible fission reaction that takes place in a nuclear reactor is shown below.

$$_{92}^{235}U + {}_0^1n \rightarrow {}_{36}^{90}X + {}_{56}^{143}Y + ...{}_0^1n$$

(a) Answer the following questions using numbers from the list. Each value may be used once, more than once, or not at all.
   235   36   2   3   90   92
   (i) Complete the equation above. [1]
   (ii) Complete the following sentences. [3]
      I   The number of protons in a uranium (U) nucleus is ..... .
      II The number of particles in a nucleus of element X is ..... .
      III The number of protons in the nucleus of another isotope of uranium is ..... .
(b) (i) Name the part of a nuclear reactor that slows down neutrons. [1]
   (ii) Name the part of a nuclear reactor that prevents an uncontrollable chain reaction. [1]

WJEC GCSE Additional Science/Physics P2 Foundation Tier January 2016 Q1

→

2 The following equation shows a nuclear reaction.

Reactants   Products

$$_1^2H + {}_1^3H \rightarrow {}_2^4He + {}_0^1n$$

(a) The reactants have to move very quickly for this reaction to take place and controlling this reaction on Earth is difficult. Complete the following sentences. [2]

  (i) The reactants are made to collide with high energies by making the gas ....

  (ii) The problem this causes is ....

(b) Select the correct word in the brackets in each sentence below. [3]

  (i) The reactants are isotopes of [hydrogen / helium / neutrons].

  (ii) The reactants have the same numbers of [neutrons / protons / nucleons].

  (iii) This reaction is an example of a [fusion / fission / chain] reaction.

(c) Give two reasons why this reaction is likely to be important in the future. [2]

WJEC GCSE Additional Science/Physics P2 Foundation Tier May 2016 Q3

3 Energy can be released in nuclear fission and nuclear fusion reactions.

(a) Explain how a sustainable, controlled chain reaction is achieved in a nuclear fission reactor containing uranium fuel rods, a moderator and control rods. [4]

(b) Explain why controlled nuclear fusion reactions are difficult to achieve on Earth. [2]

WJEC GCSE Additional Science/Physics P2 Higher Tier May 2016 Q4

4 (a) Select the correct word or phrase in brackets to complete each sentence about nuclear reactors. [3]

  (i) The function of the moderator is to [slow down the neutrons / provide channels for the cooling gas / speed up the reaction].

  (ii) The function of the control rods is to [absorb neutrons / provide channels for the cooling gas / contain the fuel rods].

  (iii) The function of the steel and concrete container is to [stop a nuclear explosion / absorb radiation / contain the plasma].

(b) The following nuclear reaction can take place in a nuclear reactor. Use the diagram in Figure 18.3 to help you answer the questions that follow.

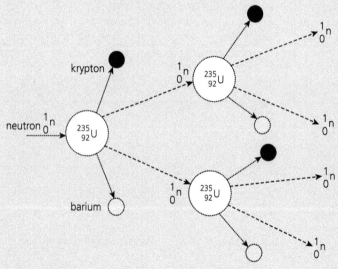

Figure 18.3

  (i) Write down the name of this type of reaction. [1]

  (ii) Name one waste product of this reaction. [1]

(c) Give two reasons why the safe storage of waste materials from nuclear reactors is controversial. [2]

WJEC GCSE Additional Science/Physics P2 Higher Tier May 2016 Q6

Answers and quick quiz 18 online

ONLINE

Now test yourself answers

Chapter 1 Electric circuits

1 In series with the other components

2 Total current = 1.3 A + 0.8 A = 2.1 A

3 If one of the bulbs failed, the other 11 bulbs would carry on working.

4
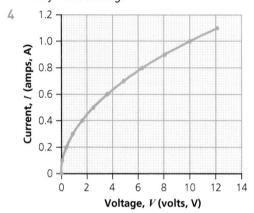

5 $I = \dfrac{V}{R} = \dfrac{3.0\,\text{V}}{15\,\Omega} = 0.2\,\text{A}$

6 $P = VI = 3.0\,\text{V} \times 0.2\,\text{A} = 0.6\,\text{W}$

Chapter 2 Generating electricity

1 Gas generation is 46 %, wind is 2 %; 23 times more electricity is generated from gas than wind.

2 Both can generate large amounts of reliable electricity.

3 Wind and wave

4 Although nuclear-power generated electricity is very expensive to commission, run and decommission, it is very reliable and can generate large amounts of electricity.

5 chemical energy (in the fuel) → kinetic energy (of the steam) → kinetic energy (of the turbine) → electrical energy (from the generator)

6 *Any two of*:
   ● to deliver a reliable and secure energy supply
   ● to match supply to demand over time
   ● to cope with sudden surges in demand.

7 To reduce energy losses

8 $P = VI = 230\,\text{V} \times 13\,\text{A} = 2990\,\text{V}$

9 $I = \dfrac{P}{V} = \dfrac{2500\,\text{W}}{230\,\text{V}} = 10.9\,\text{A}$

10
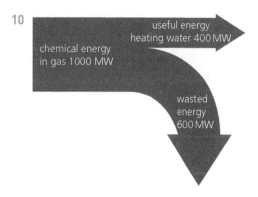

11 % efficiency $= \dfrac{400\,\text{MW}}{1000\,\text{MW}} \times 100 = 40\,\%$

12 $18\,\text{W} \times \dfrac{90}{100} = 16.2\,\text{W}$

Chapter 3 Making use of energy

1 By conduction through the glass and the plastic/metal surround of the lamp; by convection of the air around the lamp and by the emission of infrared radiation away from the hot lamp

2 Metals contain mobile electrons within their structure that conduct the thermal energy very well.

3 density $= \dfrac{\text{mass}}{\text{volume}} = 14\,400\,\text{kg}/6\,\text{m}^3 = 2400\,\text{kg/m}^3$

4 mass = density × volume = 1000 kg/m³ × 0.5 m³ = 500 kg

5 Convection

6 $54\,\text{MJ} = 54\,000\,000\,\text{J} = \left(\dfrac{54\,000\,000\,\text{J}}{3\,600\,000\,\text{J}}\right) = 15\,\text{kWh}$

7 Gas would give the cheapest running costs as it has the lowest cost per kWh.

8 annual savings $= \dfrac{\text{installation cost}}{\text{payback time}}$

   $= \dfrac{£350}{2.5} = £140$ per year

Chapter 4 Domestic electricity

1 energy transferred (J) = power (W) × time (s)
   energy transferred (J) = 200 W × (15 min × 60 s) = 180 000 J = 180 kJ

2 number of units used (kWh) = power (kW) × time (h)
   number of units = 2 kW × 11 h = 22 kWh (units)
   cost = number of units × cost per unit
   cost = 22 kWh × 15 p = £3.30

3 $P = VI$; re-arranged: $I = \dfrac{P}{V} = \dfrac{2200\,W}{230\,V} = 9.6\,A$.
A 13 A fuse would be suitable.

4 d.c. current only flows in one direction; a.c. current flows in one direction for half its cycle and then in the opposite direction for the other half of the cycle.

5 If too much current is drawn from the consumer unit, it will cut off the circuits.

6 An mcb can be reset, whereas a cartridge fuse has to be replaced.

7 $\dfrac{£3500}{£700\ per\ year} = 5\ years$

8 Wind turbines only generate electricity when the wind is blowing hard enough. Photovoltaics generate electricity during daylight.

Chapter 5 Features of waves

1 $\dfrac{50\,m}{10\,s} = 5\,m/s$

2 $f = \dfrac{v}{\lambda} = \dfrac{5\,m/s}{40\,m} = 0.125\,Hz$

3 $v = 5000\,Hz \times 0.792\,m = 3960\,m/s$

4 The direction of wave motion of a transverse wave is at right angles to the direction of vibration of the wave. The direction of wave motion of a longitudinal wave is in the same direction as the direction of vibration.

5 Ionising radiation can interact with atoms and damage cells due to their large energies.

6 (a) Radio waves

(b) Ultraviolet

(c) Infrared; microwaves

(d) All of them

(e) Radio waves; microwaves; infrared; visible light

7 angle of incidence = angle of reflection

8 An imaginary line at right angles to a mirror or boundary between mediums, from which angles of incidence, reflection or refraction are measured.

9 Water waves slow down when they travel from deep water into shallow water.

10 Microwaves travel in straight lines so a clear line of sight must be available to communicate over long distances. As the Earth's surface is curved, microwave signals can be beamed up to a satellite and transmitted to the other side of the Earth.

Chapter 6 The total internal reflection of waves

1 The beam undergoes total internal reflection.

2 One set is needed to take light from a source down into the body, and another set to transmit the reflections back up.

3 No ionising radiation is used; a biopsy can be taken as well; close-up, colour images can be taken.

4 (a) Microwaves

(b) Infrared

(c) Microwaves

5 Signals travelling down optical fibres travel at the speed of light for that material. Electrical signals travel much more slowly down copper cables.

6 The signal decreases in strength as it travels down the fibre, so it must be boosted every 30 km.

7 (a) Time delay = $\dfrac{18\,400\,000\,m}{200\,000\,000\,m/s} = 0.092\,s$

(b) Total distance = $\dfrac{76\,000\,000\,m}{300\,000\,000\,m/s} = 0.25\,s$

Chapter 7 Seismic waves

1 S-waves

2 S-waves travel slower than P-waves.

3 P-waves

4 S-waves cannot travel through the liquid outer core.

5 P-waves

6 Surface waves propagate more slowly across the surface of tectonic plates (typically with speeds of between 1 and 6 km/s).

7 A measuring instrument for detecting and recording the amplitude of earth/ground movement during an earthquake

8 Horizontal axes = time; vertical axes = amplitude of vibration

9 P-waves

10 The time interval between the arrival of the P-waves and S-waves at a measuring station.

11 Three stations widely spaced apart are needed to triangulate the epicentre of the earthquake.

12 P-travel time = $\dfrac{distance}{speed} = \dfrac{145\,km}{8\,km/s} = 18\,s$. The P-waves would be detected in Wrexham at 18 hrs : 09 mins : 30 s.

13 S-waves

14 P-waves and S-waves

15 P-waves

16 The speed of seismic waves increases with depth, causing the waves to refract and change direction.

17 The P-waves move from the mantle into the outer core.

Chapter 8 Kinetic theory

1 (a) 3.125 N/cm²

(b) 6.25 N/cm²

2 400 000 N

3 0.125 m²

4 2.25×10^5 Pa
5 2 litres
6 352 m³
7 209 litres
8 1 342 000 J
9 8736 J
10 13 475 J

Chapter 9 Electromagnetism

1 As per Figure 9.1
2 As the strength of the magnetic field increases, the magnetic field lines get closer together.
3 Turn it on and off / vary the strength of the magnetic field / reverse the field (without turning the magnet around).
4 Force
5 0.0168 N
6 23.8 A
7 Increase the strength of the magnet / more coils / higher current.
8 (a) Increases the induced voltage
   (b) Decreases the induced voltage
   (c) Increases the induced voltage
9 B to A
10 Increases the voltage / decreases the current
11 3771 turns
12 a.c. can be transformed to high voltage/low current, reducing the energy lost in the wires as heat.

Chapter 10 Distance, speed and acceleration

1 speed = $\dfrac{\text{distance}}{\text{time}} = \dfrac{200\,\text{m}}{16\,\text{s}} = 12.5\,\text{m/s}$

2 acceleration or deceleration = $\dfrac{\text{change in velocity}}{\text{time}}$
   $= \dfrac{12.5\,\text{m/s} - 0\,\text{m/s}}{5\,\text{s}}$
   $= 2.5\,\text{m/s}^2$

3
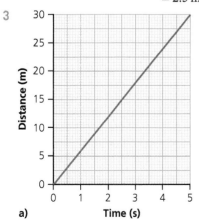
a)

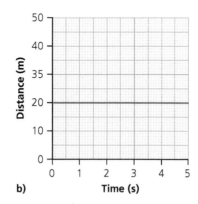
b)

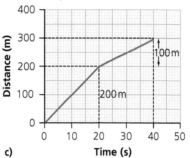
c)

4 a)
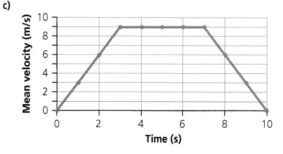

b)
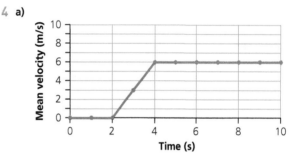

c)
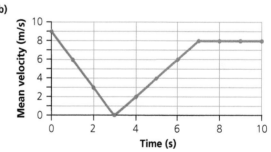

d)

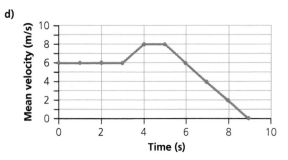

5 Thinking distance is the distance that a vehicle travels during the time it takes for a driver to decide to put on the brakes and for them to be applied. Braking distance is the distance that an object travels during the time it takes for the brakes to slow the vehicle.

6 The velocity of the car / the reaction time of the driver (which depends on tiredness, alcohol use etc.) / the driver may be distracted / the driver may hear a mobile phone ring.

7 Seat belts increase the collision time, reducing the impact force.

Chapter 11 Newton's laws

1 (a) $a = \dfrac{20\,\text{m/s}}{40\,\text{s}} = 0.5\,\text{m/s}^2$

(b) $F = ma = 0.5\,\text{m/s}^2 \times 1600\,\text{kg} = 800\,\text{N}$

2 (a) 'An object at rest stays at rest, or an object in motion stays in motion with the same speed and in the same direction, unless acted on by an unbalanced force.'

(b) resultant force, F (N) = mass, m (kg) × acceleration, a (m/s^2)

3 weight = 80 kg × 10 N/kg = 800 N

4 The weight of the skydiver acting downwards is greater than the air resistance acting upwards, producing a resultant force downwards, causing the skydiver to accelerate.

5 Eventually the force of air resistance acting upwards equals the weight of the skydiver acting downwards. There is no resultant force, hence the skydiver moves at a constant, terminal velocity.

6 'For every action force, there is an equal and opposite reaction force.'

7 (a) 200 N

(b) In the opposite direction to the force of 200 N acting on the tackled player.

(c) Contact forces

Chapter 12 Work and energy

1 PE = mgh = 0.44 kg × 10 N/kg × 20 m = 88 J

2 PE = mgh so $h = \dfrac{\text{PE}}{mg} = \dfrac{1500\,\text{J}}{(100\,\text{kg} \times 10\,\text{N/kg})} = 1.5\,\text{m}$

3 KE = $\frac{1}{2}mv^2$ = 0.5 × 80 kg × (10 N/kg)^2 = 4000 J

4 KE = $\frac{1}{2}mv^2$ so $v = \sqrt{\dfrac{2 \times \text{KE}}{m}} = \sqrt{\dfrac{2 \times 2\,\text{J}}{0.44\,\text{kg}}} = 3\,\text{m/s}$

5 $F = kx$ = 25 N/kg × 0.14 m = 3.5 N

6 work done = area under a force–extension graph

7 $W = \frac{1}{2}Fx$ = 0.5 × 3.5 N × 0.14 m = 0.245 J

8 Improving the aerodynamics of the car body / improving the aerodynamics of the car wheels / reducing energy lost when idling / using lighter-weight materials

9 Crumple zones increase the collision time, reducing the deceleration and reducing the collision force.

Chapter 13 Further motion concepts

1 $p = mv$ = 5 × 10^{-6} kg × 400 m/s = 2 × 10^{-3} kg m/s

2 $p = mv$, so $v = \dfrac{p}{m} = \dfrac{(2 \times 10^{-3}\,\text{kg m/s})}{5\,\text{kg}} = 4 \times 10^{-4}\,\text{m/s}$

3 $F = \dfrac{(53\,\text{kg m/s} - 35\,\text{kg m/s})}{3\,\text{s}} = 6\,\text{N}$

4 $\Delta p = F \times t$ = 240 N × 15 s = 3600 kg m/s

5 total momentum before an interaction = total momentum after an interaction

6 (0.25 kg × 3 m/s) = (0.75 kg × v); v = 1 m/s

7 (a) $v = u + at$ = 1.5 m/s + (0.5 m/s^2 × 3 s) = 3.0 m/s

(b) $x = ut + \frac{1}{2}at^2$

$= (1.5\,\text{m/s} \times 3\,\text{s}) + \left(0.5 \times 0.5\,\text{m/s}^2 \times (3\,\text{s})^2\right)$

$= 6.75\,\text{m}$

8 sum of clockwise moments = sum of anticlockwise moments

9 $M = Fd$ = 8 N × 0.3 m = 2.4 Nm

10 3000 N × 12 m = W × 4 m; W = 9000 N

Chapter 14 Stars and planets

1 The four inner planets are small, rocky planets and three of them have atmospheres. The four outer planets are giant planets composed of gases such as hydrogen and helium.

2 As the orbital radius increases, so does the orbital period.

3 (a) 5.2 AU

   (b) 7.8×10^{11} m

   (c) 8.2×10^{-5} ly

4 y-axis: luminosity (or absolute magnitude); x-axis: temperature (or spectral class)

5 (a) red-dwarf stars

   (b) white-dwarf stars

   (c) blue giant stars

   (d) red-giant stars

6 Main-sequence stars stretch from top left (hot, bright stars) to bottom right (cool, dim stars).

7 A supernova is the huge explosion that occurs when a massive star collapses.

8 main sequence → red giant → planetary nebula → white dwarf → red dwarf → black dwarf

9 Similarity: both formed from the collapse of supergiant stars during a supernova; differences: black holes are higher mass than neutron stars / the force of gravitational attraction inside a black hole is so large that not even light can escape.

Chapter 15 The Universe

1 'The speed of recession is proportional to the distance of the galaxy away from Earth.'

2 (a) One mark is given for a correct and relevant statement, e.g. light from sun/star passes through atmosphere of Sun/star. The second mark is only given if you correctly and coherently link a second point to the first, e.g. atoms of the gas in the atmosphere absorb light at specific wavelengths.

   (b) Galaxy 2 is further away than galaxy 1. The second mark is given for a correct relevant statement, e.g. the Universe has expanded since the light was sent out (so the waves are 'stretched') or an equivalent statement in terms of red-shift. The final mark is only given if you correctly and coherently link a third point to the second, e.g. light from galaxy 2 is red-shifted more than that from galaxy 1.

3 (a)

Element	Wavelength (nm)	Present in the star?
Helium	447, 502	Y
Iron	431, 467, 496, 527	N
Hydrogen	410, 434, 486, 656	Y
Sodium	590	Y

   (b) The wavelengths of the lines would be greater / the position of the lines would be shifted towards the long-wavelength end of the spectrum (or red-shifted / shifted to right / shifted to red); because distant galaxies are moving away (from us) / because of the expansion of the Universe/space.

4 Cosmological redshift shows us that the rate of expansion of the Universe is increasing; that the expansion of the Universe is 'accelerating'. Working backwards gives evidence for the Big Bang.

5 Since the Big Bang, the wavelength of the gamma rays emitted at the time have been stretched so much that the background remnant of these gamma rays now have the wavelength of microwaves.

Chapter 16 Types of radiation

1 $^{7}_{3}$Li; $^{13}_{6}$C; $^{90}_{56}$Sr; $^{99}_{43}$Tc

Type of radiation	Symbol	$^A_Z X$	Penetrating power	Ionising power
Alpha	α	$^4_2 He$	Stopped by sheet of paper	High
Beta	β	$^0_{-1} e$	Stopped by a few mm of Al	Medium
Gamma	γ		Reduced by several cm of Pb	Low

3 (a) A = 237; Z = 93

 (b) A = 225; Z = 89

4 Background radiation is all around us and it comes naturally from our environment and from artificial (human-made) sources.

5 Repeat measurements and take a mean average; subtract background; long count rates.

6 (a) Paper and aluminium have no effect on the mean count rate.

 (b) B, because it is the only source where count rate is affected by a piece of paper.

 (c) (i) Beta radiation (or electron) emitted from the nucleus / to produce a stable nucleus.

 (ii) $^{90}_{38}Sr \rightarrow ^0_{-1}e + ^{90}_{39}Y$; A = 90; Z = 39; element = Y (yttrium)

Chapter 17 Half-life

1 Radioactive decay is a random process but it obeys the rules of probability, and each radioactive atom has a given probability of decay. All radioactive substances decay in the same way, but with different time frames.

2 Half-life is the time taken for the activity of a radioactive substance to halve in value.

3 (a) (i) 6 hours

 (ii) 6 hours

 (iii) 12 hours

 (b) Alpha is absorbed easily / would not be detected outside the body. It is highly ionising / causes damage to DNA in cells/tissues/organs.

4 Half-life; penetrating power; ionising ability

5 Radio-imaging; radiotherapy

6 (a) Correctly determined points: after 5700 years activity = 32 counts/minute, after 11 400 years activity = 16 counts/minute, after 17 100 years activity = 8 counts/minute; make sure you plot the points accurately and draw a smooth best-fit line.

 (b) (i) C-14 begins to decay

 (ii) 4500 ± 300 years with lines shown on the graph or a written explanation.

Chapter 18 Nuclear decay and nuclear energy

1 Nuclear fission is the breaking apart of unstable nuclei.

2 $\left(\dfrac{83\,000\,000\,000\,000\,J}{35\,000\,000\,J} \right) = 2\,371\,429$ times (approximately 2.4 million)

3 (a) Atomic number of Ba = 56; mass number of Kr = 89

 (b) (i) Slows down (fast) neutrons

 (ii) Boron (steel) rods raised from / lifted from the reactor / add more fuel/uranium / increase the number of successful collisions / absorb fewer neutrons.

4 A nuclear reaction involving the fusion (joining together) of light nuclei.

5 Nuclear fusion reactions produce huge numbers of high-energy neutrons that need shielding from the surroundings.

6 (a) (i) Hydrogen

 (ii) Fusion

 (b) High temperature/energy is needed (for particles to overcome their repulsive force), but that would melt the container. High pressure is needed, so needs to be very strong containment.

 (c) Greater availability of fuel; waste material is not / is less radioactive; more energy is available from fusion than fission.